UNDERSTANDING
SUPERCOMPUTING

UNDERSTANDING SUPERCOMPUTING

FROM THE EDITORS OF *SCIENTIFIC AMERICAN*

Compiled and with introductions by

Sandy Fritz

Foreword by David A. Patterson

A Byron Preiss Book

WARNER BOOKS

An AOL Time Warner Company

If you purchase this book without a cover, you should be aware that this book may have been stolen property and reported as "unsold and destroyed" to the publisher. In such case neither the author nor the publisher has received any payment for this "stripped book."

Copyright © 2002 by Scientific American, Inc., and
Byron Preiss Visual Publications, Inc.
All rights reserved.

The essays in this book first appeared in the pages of *Scientific American*, as follows: "Blitzing Bits" by W. Wayt Gibbs, *Scientific American Presents Extreme Engineering*, Winter 1999. "The Law of More," by W. Wayt Gibbs, *Scientific American: The Solid-State Century* 1997. "Getting More from Moore's" by Gary Stix, April 2001. "The Future of the Transistor" by Robert W. Keyes, *The Solid-State Century* 1997. "Microprocessors in 2020" by David A. Patterson, *The Solid-State Century* 1997. "A Vertical Leap for Microchips" by Thomas H. Lee, January 2002. "The Semiconducting Menagerie" by Ivan Amato, *The Solid-State Century* 1997. "Engineering Microscopic Machines" by Kaigham J. Gabriel, September 1995. "From Chips to Cubes" by W. Wayt Gibbs, November 1997. "Quantum-Mechanical Computers" by Seth Lloyd, *The Solid-State Century* 1997. "Quantum Computing with Molecules" by Neil Gershenfeld and Isaac L. Chuang, June 1998. "Computing with Light" by Graham P. Collins, August 2001. "Double Quantum Dot Computing" by Kristin Leutwyler, September 2001. "The Do-It-Yourself Supercomputer" by William W. Hargrove, Forrest M. Hoffman and Thomas Sterling, August 2001. "Tackling Turbulence with Supercomputers" by Parviz Moin and John Kim, January 1997. "How to Build a Hypercomputer" by Thomas Sterling, July 2001.

Warner Books, Inc., 1271 Avenue of the Americas, New York, NY 10020

Visit our Web site at www.twbookmark.com.

Printed in the United States of America
First Printing: December 2002
10 9 8 7 6 5 4 3 2 1

ISBN: 0-446-67957-7
Library of Congress Control Number: 2002108952

🅦 An AOL Time Warner Company

Cover design by j. vita
Book design by Gilda Hannah

ATTENTION: SCHOOLS AND CORPORATIONS
WARNER books are available at quantity discounts with bulk purchase for educational business or sales promotional use. For information, please write to SPECIAL SALES DEPARTMENT, WARNER BOOKS, 1271 AVENUE OF THE AMERICAS, NEW YORK, NY 10020.

Contents

ated into our environment. T
everywhere-in grocery check
nes, inside the workings of
meras, under the hoods of ca
ey direct the functioning of
tches, artificial hearts, an
tomotive assembly lines. Som
wly designed fighter jets ar
stable that without constant

Foreword

David A. Patterson

Supercomputers are crystal balls of legend, allowing us instantaneous travel in time and space. They let us travel back to the Big Bang, see subatomic particles that the human eye should never be able to see, and watch what global warming might do to our planet. This book will introduce you to the many alternative futures of supercomputing, from the fantastic to the mundane, and the challenges and pitfalls inherent in realizing them.

Today's supercomputers are composed of the same silicon chips found in desktop computers; supercomputers just use thousands of them. Silicon is the miracle material that Gordon Moore predicted would be able to double the computing resources per chip every 18 months. Moore's Law has been sustained for 30 years; how much longer it can go is a matter of considerable speculation by engineers, scientists, and entrepreneurs.

As long as Moore's Law remains valid, I expect supercomputers will be built from silicon chips based on principles known for decades. People who adhere to this mainstream view of supercomputers place their faith in parallelism. If we connect thousands to millions of these chips together, the giant problems that we face can be solved by working on many parts of the problem simultaneously. We depend upon inventors to find new solutions that let problems be solved in parallel, and on programmers to transform such inventions into code that runs well on such a coordinated legion of computers.

But there are other futures, ones that you might expect more likely to be found in science fiction than in the pages of *Scientific American*. Serious scientists are spending their lives attempting to validate this countercultural approach to silicon supercomputing.

Despite such efforts, be forewarned that many fascinating technologies have been crushed by Moore's Law and the silicon juggernaut.

Micro Electrical-Mechanical Systems (MEMS) and nanotechnology are miniature mechanical devices built on silicon. Many storage devices today are mechanical, such as magnetic disks. Hence, the most obvious use of MEMS is for storage, offering a tiny alternative to such devices that maintain information. Engineers are also exploring computation based on mechanical devices.

The next step in our fantastic voyage is computing with DNA. Suppose we consider strands of DNA to be strings of information. Perhaps the instruction that transforms information in our biological computers could be a reproductive process that produces a complementary DNA strand in the presence of an enzyme. After placing the chemicals in the test tube, you wait until the molecules are born with the right answers encoded within them, in part by adding more chemicals to weed out the wrong answers.

Beyond micromechanical and biological computers is quantum computing. Exploiting the quantum affects of physics that in a single step could theoretically calculate all possible answers, these magical devices are the ultimate parallel computers. Some computer scientists today are inventing algorithms assuming quantum computers will appear, placing their faith in future computer designers to turn quantum computing into reality. Thus far it appears there is a devil's compact that promises phenomenal computation rates, but only for a second or so before the quantum machines break down into random information. As a result, quantum computer architects are looking at designs that spend most of their resources in correcting errors to counteract this tendency to "decohere."

Another future vision is that of optical computers. Optics has already proved itself in transmission via optical fibers; the question is whether optics can be used to construct the guts of microprocessors, so that photons would replace electrons in storage and computation as well as in transmission. As Moore's Law continues, an increasing challenge is the time and power for conventional designs to transmit information across relatively large silicon chips using tiny transistors. Distance and power are much smaller problems for light, which is a reason for optical computer designers to be optimists. By using the wave nature of light, it may even be possible to build the optical equivalent of a quantum computer.

The final fantastic material for a supercomputer is . . . really nothing at all. It reimagines the millions of idle PCs connected to the Internet as a sleeping supercomputer, to be awakened by the proper incantation. The first step is finding a problem that can be broken into miniscule pieces that compute for a haphazard amount of time for which the individual machines rarely need to communicate. The group working on the Search for ExtraTerrestrial Intelligence (SETI) have such a problem, looking for signs of intelligent life from signals of billions of stars. The next step in awakening a sleeping supercomputer is getting permission from millions of people to use their idle computers. SETI relied on charity, hoping that the love of science and the possibility of contributing to a breakthrough would lure millions to run pieces of the problem. SETI's success has led to start-up companies trying the same tactic. The challenge is finding similarly partitionable problems that someone will pay to solve while rewarding PC owners and still making a profit.

We cannot know whether these fantastic devices will become the mainstream of computing, find a role for a limited problem domain, replace one component of a standard computer, or, if Moore's Law continues, simply be vanquished as a pretender to silicon's throne.

We may not know the ultimate answer, but the likely alternatives are presented here, letting the reader see the future of supercomputing no matter what the outcome.

Introduction

Sandy Fritz

Microchips have been seamlessly integrated into our daily environment. They're everywhere—in grocery check-out lines, inside the workings of cameras, under the hoods of cars. They direct the functioning of watches, artificial hearts, and automotive assembly lines. Some newly designed fighter jets are so unstable that without constant adjustments from a bank of on-board computers powered by microchips, they couldn't fly. The space shuttle cannot be flown manually at all.

When bank after bank of microchips are laced together and fitted to a single motherboard, the supercomputer is born, a single machine that uses the sheer strength of its computing power to muscle through problems. And like its distant relative the humble PC, silicon microchips are at the heart of these superfast computing devices.

Silicon chips gave birth to the modern computer, but they also represent a barrier for ongoing computer growth. Microchip circuitry is becoming so small that manufacturers are quickly reaching the physical limit for how small they can craft a silicon microchip that will function. Other materials are under investigation as possible replacements, but there's a good chance that by 2015, the world's smallest possible silicon microchip could force changes in how computers operate.

One solution to the physical speed limits of a single or small cluster of microchips is to have many computers working on small parts of a single problem at the same time. This approach, called parallel computing, has been extensively used in cash-strapped research centers to cobble together a network of computers whose speed and function come close to super-computer performance but at a fraction of the cost.

The Internet itself represents a potent probable supercomputer waiting to be awakened. Through parallel computing, tasks could be parceled between millions of machines, vastly speeding complex processing time. This "virtual supercom-puter" could have nodes literally all over the world.

As the problems of physical speed and size limits for silicon chips grow nearer, another camp of thought is taking a whole new look at what computers are, and pondering new ways in which they might be designed. Nanotechnology innovations may someday be able to assemble computer chips atom by atom. Molecules themselves might function as switches and transistors. By opening the science to this kind of creative thinking, new ideas and insights are generated.

The ever-shrinking size of microchips is taking them close to the quantum mechanics twilight zone, and it is in this counter-intuitive yet pervasively influential realm of matter that some of the most interesting thinking about computing is taking place. In theory, a quantum computer could be vastly faster than any digital computer, and its CPU might resemble a pill filled with water. But getting the theory that works on paper to work in the lab has been problematic.

The next technological advancement on the supercomputer may be a machine called the hypercomputer. If built, the hypercomputer would outstrip the speeds of all existing computers. On the edge of these world's fastest machines some of science's most important questions are asked. As supercomputers become more powerful, they are able to help us answer these questions.

Three key approaches shape the designs of supercomputers, each with its own virtues and its own shortcomings. One approach is to assemble a powerhouse of chips under a single housing that, through sheer processing power, tackles problems. A cheaper alternative to the all-purpose supercomputer is a machine that is custom designed to work with certain problems. Finally, and perhaps most intriguing, is a virtual supercomputer formed of millions of common PCs linked via the Internet.

Blitzing Bits

W. Wayt Gibbs

S troll into your local computer store, plunk down $2,000, and you can take home a fairly zippy machine. With a seventh-generation 600-megahertz processor, 128 megabytes of memory and about 13 gigabytes of disk capacity, a late-model personal computer can tear through that full-screen, full-motion, post-apocalyptic shoot-'em-up with scarcely a hiccup. But snazzy computer games are one thing. Simulating what actually happens inside a detonating nuclear bomb—or a collapsing star or a folding protein—requires a qualitatively different kind of machine, a machine that has not yet been built. It took the Blue Pacific system at Lawrence Livermore National Laboratory, currently the fastest supercomputer in the world, 173 hours to complete a simple turbulence simulation. Your state-of-the-art PC would have to hum along for well over 16 years to do the same job, assuming it worked at its peak speed of 600 megaflops (million floating-point operations per second)—which, of course, computers never do. Every 20 days you would have to add another 13-gigabyte hard drive to store the results.

And yet, says Mark K. Seager, one of the supercomputer gurus at Livermore, this massive computation will shed light on just one small, idealized part of the problem. To confidently answer whether refurbished bombs will burst, how the globe will warm, how the universe took shape, and other questions that elude theory and experimentation, scientists need computers of 10 to 10,000 times the speed and capacity of Blue Pacific. One such machine is already under construction, and others are on the drawing board.

Three Ways to Get from Here to Petaflops

For any other field of engineering, the idea that one could achieve a thousandfold increase in performance in less than a decade would be sheer lunacy. We will not soon see cars streaking along the desert at 700,000 miles (1.1 million kilometers) per hour or buildings that rise 280 miles into outer space. But many computer scientists believe that by 2007 they will be able to build a supercomputer that delivers one petaflops (a quadrillion floating-point operations per second)—three orders of magnitude faster than Blue Pacific.

If it were made in the mold of ASCI White, a petaflops machine would contain at least 250,000 microprocessors, draw one billion watts—the output of a large nuclear power plant—and cost roughly as much as a fleet of aircraft carriers, estimates Thomas L. Sterling of the California Institute of Technology. Its processors would waste most of their time waiting for data to arrive from memory. To skirt these obstacles, researchers are investigating at least three radically different designs.

1. A consortium led by Caltech is working on the most ambitious of the three approaches to petaflops capability, a so-called

hybrid-technology multithreading architecture. "Hybrid technology" means that the researchers intend to use new kinds of chips, networks, disks—everything. Massive microchips will have logic circuits woven among large banks of memory, so that the two can communicate more rapidly. The "single quantum flux" chips will be cooled to near absolute zero so that they superconduct and use about one millionth the energy of conventional processors. This also should allow them to run at speeds exceeding 150 gigahertz, so that "only" 2,048 processors are needed for the system's multithreaded operation, in which a program is broken into individual tasks that can be performed concurrently.

Information will flow through the system as light in optical fibers rather than as electricity in copper cables, increasing bandwidth by a factor of 100 or more. And up to one petabyte (million billion bytes) of data will be stored as holograms in crystals rather than as magnetic patterns on spinning disks, greatly boosting speed and reducing power consumption.

The consortium has prototypes of some of the components, but a number of the technologies are still in the research stage. Nevertheless, with backing from four federal agencies, this effort is the best funded of the petaflops supercomputer designs.

2. The first computer to break the teraflops barrier was no bigger than a large photocopy machine, and it sat in a humble room at the University of Tokyo. Called GRAPE-4, it did only one thing—calculate the gravitational attractions among many objects such as stars or asteroids—but it performed its job with exceptional efficiency, surpassing on that narrow range of problems the speed of even the mighty Blue Pacific.

GRAPE-6, now nearing completion, should set another milestone, hitting 200 teraflops by executing sizable chunks of its program on special-purpose microprocessors. This is a cheap way to build supercomputers if you need them for just one kind of problem, says Mark Snir, manager of scalable par-

Inside the Fastest Computer

Blue Pacific, the U.S. Department of Energy's 3.9-teraflops (trillion floating-point operations per second) supercomputer, became fully operational in May 1999. Just over one year later it will be made obsolete by the next step in the DOE's Accelerated Strategic Computing Initiative (ASCI). In a giant lab in Poughkeepsie, N.Y., IMB engineers are constructing a $100-million, 10-teraflops successor, called ASCI White, which will occupy a large room *(below)* and is scheduled for demonstration in March 2000 and for operation by late summer. Like Blue Pacific, the new machine will divide up programs to run on thousands of processors simultaneously.

ASCI White is but the fourth of seven supercomputers planned by the initiative, which aims to produce a 100-teraflops system by 2004. That level of performance, says Livermore's Mark Seager, is "the absolute minimum" needed to simulate how an entire nuclear weapon would detonate—or not.

1 In January 1997 the sun expelled a wave of plasma that collided with Earth's magnetosphere, creating stunning auroras and providing a rich set of scientific observations that may help explain how our planet's magnetic shield works. To compare the observations with current theory, physicists at the University of Maryland created a detailed supercomputer simulation of the event, computing a kind of three-dimensional movie from the basic laws of physics.

2 On ASCI White, such a simulation would first carve out an imaginary block of space around Earth. In order to perform the calculations as quickly as possible, the software would divide this volume *(above)* into perhaps 10 billion smaller "cubes." These units, each containing only a few mathematical operations and some initial numbers, would stream out of random-access memory (RAM) and move through a switch.

3 The heart of any supercomputer is the switch that pulses data between and among processors, memory chips and disks. ASCI White uses a "multistar omega network," which connects the 8,192 processors in the machine with one another and with 10,752 external disk drives in such a way that any processor is never more than two hops away from any other one. The switch can move data to and from each group of processors at a rate of 800 megabytes per second—more than five times the speed of Blue Pacific's switch.

# of Tomorrow

6 After a week or two of around-the-clock operation, the final movie of 50,000-frames—a total of perhaps 500 trillion cubes—will be complete (*right*). To store such massive amounts of data, ASCI White will boast 195 terabytes of external disk storage (*left*). By way of comparison, the printed contents of the Library of Congress comprise about 10 terabytes.

5 As the cubes of information enter the nodes, they flow into memory and are distributed among the processors. If all the numbers are in place, the processor can do its mathematical work, filling the cube (*left*) with the results and sending it back out over the switch to be stored in the disk farm. Often, however, the data in a cube are in RAM or are spread among two or more processors that must pass messages to one another to cooperate in arriving at an answer. This process slows the computation enormously, and as a result supercomputers rarely operate at more than 20 percent of their theoretical peak speed.

4 The processors are organized into nodes, which are grouped four to a case. Every node (*left*) in turn houses 16 375-megahertz POWER 3-II microprocessors. This chip is designed to execute four floating-point calculations simultaneously, for a peak performance more than twice that of a 600-megahertz Pentium III. In addition to more than eight megabytes of cache RAM per processor, every node contains at least eight gigabytes of local memory and two internal 18-gigabyte hard drives.

allel systems at IBM. "We looked here at what it would take to build a multipetaflops machine customized to the problem of protein folding," he says. "We could do it for a few million dollars—much, much less than a general-purpose machine."

There may be a way to have the best of both worlds. Recent generations of so-called configurable chips—processors that can rewire their circuitry on the fly—raise the hope of supercomputers that can transform themselves into ultrafast machines custom-designed for the problem at hand. But configurable chips are still so slow and expensive that the idea remains little more than a hope.

3. The fastest and hardest-working computer system on the planet is not behind razor wire at a classified lab or humming in the bowels of some university building. It cost less than $1 million to set up and almost nothing to run. It never had to come down for maintenance, and it grew faster every day.

SETI@home, a small program written at the University of California at Berkeley and distributed over the Internet, was released in 1999. Within three months, more than a million people had downloaded the software, which scans signals recorded by the Arecibo radio telescope in Puerto Rico for signs of extraterrestrial intelligence. With SETI@home installed, each PC downloads from Berkeley a chunk of data to process, performs the calculations while the machine would otherwise be idle and then sends the results back.

By September the results were pouring in at the rate of seven teraflops. Put another way, a popular screensaver had in four months zipped through computations that would have taken the Blue Pacific supercomputer about 26. This may be a special case, points out Dan Werthimer, the project's chief scientist. "I don't think we could attract one million people in 224 countries to help with one of the 'grand challenge' problems," such as turbulent mixing.

But at Berkeley and elsewhere, SETI@home does provide inspiration to researchers who are trying to build "virtual super-computers" by connecting, say, all the computers in a university or a hospital and harnessing processing power that would otherwise go to waste. That research raises the possibility that one day in the near future, the Internet will offer a way not merely to communicate but also to tap into a nearly unlimited reservoir of computing power.

The very heart of a computer—its microchip—was first commercialized in 1971 by engineer Gordon Moore. Moore established rigorous expectations for silicon-based microchips, but even he realized that there was a speed barrier inherent in the use of silicon.

roChips have seamlessly in
ated into our environment. Th
everywhere-in grocery check
nes, inside the workings of
neras, under the hoods of car
ey direct the functioning of
ches, artificial hearts, and
comotive assembly lines. Some
vly designed fighter jets are
stable that without constant

The Law of More

W. Wayt Gibbs

Technologists are given to public displays of unbridled enthusiasm about the prospects of their inventions. So a reader flipping through the 35th anniversary issue of *Electronics* in April 1965 might easily have dismissed an article by Gordon E. Moore, then head of research at Fairchild Semiconductor, pitching the future of his business. Moore observed that the most cost-effective integrated circuits had roughly doubled in complexity each year since 1959; they now contained a whopping 50 transistors per chip. At that rate, he projected, microchips would contain 65,000 components by 1975, at only a modest increase in price. "Integrated circuits," Moore wrote, "will lead to such wonders as home computers— or at least terminals connected to a central computer—automatic controls for automobiles, and personal portable communications equipment."

Technically, Moore was overoptimistic: 65,000-transistor chips did not appear until 1981. But his fundamental insight— that continued geometric growth in the complexity of microelectronics would be not only feasible but also profitable—held true for so long that others began referring to it as Moore's Law. Today, from his vantage point as chairman emeritus of Intel, Moore observes that his prediction "has become a self-fulfilling prophecy. [Chipmakers] know they have to stay on that curve to remain competitive, so they put the effort in to make it happen."

That effort grows with each generation—in 1997, Intel and its peers spent about $20 billion on research. Moore expects the rule of his law to end within the next decade, coinciding nicely with the twilight of his career. Such good fortune—the kind that tends to smile on the prepared—is a recurrent theme in the history of Moore and the microprocessor.

Even Moore's entry into the semiconductor business was accidental. A year after finishing his doctorate at the California Institute of Technology in 1954, the physical chemist decided to take a job as an inspector of nuclear explosions at Lawrence Livermore National Laboratory. By coincidence, William Shockley, one of the inventors of the transistor, was at the time looking for chemists to work in his semiconductor company and got permission to rifle through Livermore's résumé file. "I had no background whatsoever in semiconductors," Moore recalls. Shockley offered him a job anyway.

"Shockley was a technical genius, but he really didn't understand how people worked very well," Moore says. Within a year he, Robert N. Noyce and several colleagues abandoned Shockley to found a new firm. Fairchild Semiconductor produced the first commercial integrated circuit in 1959 and grew over the next decade into a $150-million business. But soon after it was bought out by a conglomerate, Moore grew restive. In 1968 he and Noyce struck out again on their own.

Fairchild and its competitors were still customizing chips for every system. "The idea we had for Intel," Moore says, was "to make something complex and sell it for all kinds of digital applications": first memory chips, then calculators. "But we were a little late," Moore adds. All the big calculator companies already had partners.

Noyce tracked down a small Japanese start-up named Busicom that had designed the logic for 13 microcircuits to go into its scientific calculators. "To do 13 different complex custom circuits was far beyond what we could tackle," Moore recounts. But after some thought, Intel engineer Ted Hoff concluded that a single, general-purpose chip could perform all 13 functions and more.

And so, out of chance and desperate necessity, the microprocessor was born in 1971. Under Moore's direction, four Intel engineers created, in nine months, a computer on a chip. There was just one problem, Moore admits, with a sheepish grin: "Busicom paid a portion of the development costs and therefore owned the rights to the design."

But fortune was again on Moore's side. Busicom slipped into financial straits. "We essentially gave them back the $65,000 [they had paid Intel] and got the rights to the chips back for all uses. So the Japanese initially owned all the rights to microprocessors but sold them for 65 grand. In retrospect, it was kind of like the purchase of Manhattan" island for $24 in 1626, he laughs.

With Moore as chief executive and later as chairman, Intel rode the wave of Moore's Law for over 25 years. But that wave will begin to break as the costs of cramming more transistors on a slice of silicon overwhelm the benefits. "Things we used to do relatively casually to advance the technology now take teams of Ph.D.'s," he says, estimating that 400 engineers worked for four years to produce Intel's latest processors.

Moore predicts that tweaking the lenses, robots and ultraviolet lasers used to etch circuits onto silicon will extract perhaps two more generations of processors, with features 0.18, then 0.13 micron across, from current optical techniques. "Beyond that, life gets very interesting," he notes. "We have three equally unattractive alternatives."

X-rays, with their smaller wavelength, could carve out wires just a handful of atoms across. But blocking such energetic waves requires very thick stencils as tiny as the chip itself. "It is very hard to make the mask perfect enough and then to do the precision alignment," Moore warns. "So while a lot of work continues on x-rays, some of us have lost our enthusiasm for that technology."

A second option is to use electron beams to draw circuit designs line by line onto silicon. But that process is still far too slow for mass production, Moore says: "And as you go to smaller dimensions, the total distance the beam has to travel to make the pattern keeps going up." Experiments with wider beams look promising, however. "Worst case, we will be able to make a layer or two of some very fine structures with an electron beam, then add optically some structures that are not so fine," he wagers.

The smart money, Moore says, is on soft (relatively low frequency) x-rays. "There is still a tremendous amount of engineering involved in making this work," he cautions. "You have to have a reflective mask instead of a transparent mask. You have to have a vacuum system. You have to have new resist [coatings]." But if it succeeds, Moore concludes, soft x-ray lithography "will take us as far as the material will let us go, a long ways from where it is now."

Moore worries as much about the consequences of tinier transistors as about ways to make them. With the rapid increase in chips' complexity and clock speeds, "if you don't do anything else, the power goes up something like 40-fold" every two generations, he points out. "Well, if you start with a 10-watt device [such as the Pentium] and go up 40-fold, the darn thing smokes! We've handled it to date by lowering the voltage. But you can only go so far on that."

To squeeze out another decade of geometric performance growth, chip manufacturers will try various tricks, Moore predicts. "Phase-shift masks [that compensate for the diffraction of laser light] allow you to go to smaller dimensions with a given wavelength." Processors are already five or six layers high; they will thicken still more. And silicon wafers will grow from eight to 12 inches in diameter, enabling greater economies of scale. Until recently, Moore concedes, "I was a skeptic there. I argued that the cost of material was going to be prohibitive." But he failed to foresee advances in crystal-growth techniques.

Moore's vision is also less than clear on what new workaday jobs will require desktop supercomputers with processors that are 10 to 100 times more powerful than today's. "We haven't identified any very general ones yet," he admits. Indeed, by the time computer hardware begins to lag Moore's vision, engineers may find that the barriers to more intelligent, useful machines lie not in physics but in software, which obeys no such law.

To take silicon-based microchips to the very limit of their ability to function, Intel initiated the process of refining extreme ultraviolet lithography, a manufacturing technique that can render transistor elements just 40 atoms in width. But at these scales, even the tiniest flaw means catastrophic failure.

ated into our environment. T
everywhere-in grocery check
nes, inside the workings of
neras, of ca
ey dire the functioning of
tches, artificial hearts, an
tomotive assembly lines. Som
wly designed fighter jets ar
stable that without constant

Getting More from Moore's

Gary Stix

When Gordon Moore, one of the founders of Intel, plotted a growth curve in 1965 that showed the number of transistors on a microchip doubling every 18 months, no one had any idea that his speculations would not just prove prescient but would become a dictate—the law by which the industry lives or dies.

Like a drug addict in search of a fix, the semiconductor industry can keep on the curve of Moore's Law only by constantly adopting new technology that requires ever greater infusions of capital and technical sophistication. Intel, the company that has served as the standard bearer for Moore's Law, has waged a five-year crusade to develop a method of printing circuit patterns on chips that could take the reigning CMOS chip technology until circuits can be made no smaller, the last data point on the Moore curve.

These new lithographic machines for making billion-transistor microprocessors will mark one of the most spectacular forays into the realm of nanotechnology, the precise manipulation of matter at the scale of a few billionths of a meter. The Intel-nurtured technology—extreme ultraviolet lithography (EUV)—has recently created one of its first images of a whole chip at a Department of Energy laboratory set up to engineer nuclear weapons. At a wavelength of 13 nanometers, EUV will eventually have the ability to print a transistor element just 40 atoms in width.

Progress toward what the industry calls its next-generation lithography lends credence to Intel's strategy of relying on collaborations with universities or national laboratories to tap a wellspring of basic research and development resources. The Intel approach stands in marked contrast to the large centralized laboratories built by AT&T, IBM and Xerox, which have often invented technologies that they never succeeded in commercializing. "The classic research model never worked," says G. Dan Hutcheson of VLSI Research, a market research firm that has tracked these technologies for 25 years. "Intel looked at research in a new way and showed how to get a return on investment from it." Even before the founding of Intel in 1968, Gordon Moore had developed a bias against the traditional approach after he witnessed Fairchild Semiconductor squandering capital on research that never turned into products during his tenure there in the 1960s.

Recent experience bolsters Intel's case. The demonstration at Sandia National Laboratories/California in Livermore comes a year or so after the demise of a lithography program, championed by IBM for decades, that used x-ray radiation. The program consumed hundreds of millions of dollars in expenditures by both IBM and the Defense Advanced Research Projects Agency—and some industry observers estimate that the sum exceeded $1 billion. Moreover, in 2001 two major semiconductor equipment manufacturers—ASML and Applied Materials—dropped plans to develop electron projection lithography, which uses parallel beams of electrons to print circuit patterns, another contender for the next-generation lithography that had been under development for years inside AT&T Bell Laboratories.

Despite its role as lead sponsor for EUV, Intel cannot claim credit for inventing it. In the late 1980s AT&T Bell Laboratories (now part of Lucent Technologies) and NTT Communications published separate papers on soft x-ray projection lithography. Two national laboratories—Sandia and Lawrence Livermore—expanded on this work using technologies from the Strategic Defense Initiative. Sandia fashioned an early lithography prototype using radiation from a laser-generated plasma, which had been involved before in testing the response of different materials to the high-energy pulses that satellites might sustain in scenarios postulated by "Star Wars" planners.

It has been understood for decades that the billion-dollar expense and overwhelming difficulties of producing chips with nanoscale circuitry would require that chipmakers such as IBM, Intel or (at one time) AT&T fund the early research of their equipment manufacturers. Bell Labs, which oversaw parallel efforts in five separate lithography technologies during the early 1990s, was enticed by the idea of short-wavelength radiation that did not require a synchrotron, the giant x-ray generators found in high-energy physics laboratories. The technical difficulties that beset x-ray lithography at the time led the Bell Labs researchers to change the name from soft x-rays to extreme ultraviolet lithography. Intel had joined AT&T and others in a cooperative research program with the national laboratories. But the actual day-to-day research was concentrated at Lawrence Livermore, Sandia and Bell Laboratories.

When Congress eliminated the program in 1996, pegging it as a form of corporate welfare, AT&T decided to get out. Intel then stepped in to salvage and carry on the work. "Intel came to the realization [that] if they didn't put money into a couple of key technologies that would come into play in the 2000s, they were going to be in big trouble," says Richard R. Freeman, a professor of applied science at the University of California at Davis, who headed lithography development at AT&T Bell Labs and later the EUV program at the national laboratories during the mid-1990s.

On paper, EUV was attractive. With a wavelength of 13 nanometers—almost one twentieth the wavelength being readied for use in commercial chipmaking five years ago—EUV could be extended until the physical challenges of making atomic-scale chips rendered existing semiconductor technologies unworkable. And the technology used a machine tool that bears some resemblance to those deployed in existing fabrication facilities. Insiders at Intel were suspicious, though. "People started asking, 'Can it do this, can it do that?,' and it was Gordon Moore who said we really don't have an alternative," recalls John Carruthers, who headed advanced technology research at Intel at the time.

Intel entered into a three-year contract (later extended to five) with an entity called the Virtual National Laboratory (VNL), which combined researchers and facilities from Lawrence Livermore, Sandia and Lawrence Berkeley national laboratories. Having one contract with three labs cut some of the red tape that usually discourages companies from seeking such collaborations. Later Intel brought in other chip manufacturers, including competitors AMD, Motorola, Micron and Infineon—and lithography equipment suppliers ASML and SVG.

In 1997, at the beginning of Intel's stepped-up involvement, looming technical difficulties caused EUV to be rated last out of four lithography technologies in a straw vote taken at an industry conference. Although it bears some similarities to existing methods, EUV is different enough to make the average fabrication-line manager quake. Conventional photolithography equipment projects ultraviolet light (usually at 248 or 193 nanometers) through a mask—a sheet of glass on which are traced a chip's circuit patterns. A series of lenses reduces the image to a quarter of its size. The image projected through the lenses is exposed in a chemical on the wafer. Another chemical treatment then etches away either the exposed or unexposed areas of the image, carving the circuit elements into the chip surface.

Things change at 13 nanometers, where extreme ultraviolet lithography earns its name. The mask and lenses, transparent at longer wavelengths, would absorb this radiation. So EUV uses mirrors for both the mask and the lenses. A laser trained on a jet of xenon gas creates a plasma that emits 13-nanometer radiation, which is focused onto a mask. The mask reflects the circuit pattern onto a series of curved mirrors that reduce the size of the image and focus it onto the wafer. The 80 alternating layers of silicon and molybdenum that make up the mirrors and the mask have to be smoothed to single-atom tolerances. The entire circuit-printing process, moreover, has to be done in a vacuum because air itself absorbs radiation at this wavelength. And the mask will distort the image if it contains more than a handful of defects measuring even 50 nanometers, about 2,000 times narrower than the width of a human hair. The development team sometimes muses on ways to describe to the outside world the relative size of a 50-nanometer defect, comparing it to a search for a golf ball in a state the size of Maryland, a basketball in the state of Texas or a hair on a football field.

Physicists and engineers who designed and engineered nuclear weapons technology had to solve these challenges. Unlike AT&T, which conducted early development work on EUV at Bell Labs with about 30 employees, Intel has only five full-time employees at VNL's main facility at the Sandia laboratory in Livermore (although more than 10 others labor on developing proprietary mask designs and other EUV-related technology at several Intel facilities). "They're using us as an advanced development and research lab," says Richard H. Stulen, the virtual laboratory's chief operating officer.

The company kept a close eye on how decisions were made at the labs. If alternative methods were proposed for making lenses, Intel would press the research team to pick one, instead of testing the merits of both. "Nothing got spent that they didn't think would work," Freeman says. "They didn't do it Bell Labs style." Intel also implemented the same detailed risk-management system that the company uses internally—essentially a rating system of things that could go wrong. This flagged a list of about 200 problems, some of which the 150 national laboratories researchers who worked in the VNL might otherwise have downplayed. At one meeting, the VNL staff mentioned that it would need to increase the power of the laser by a factor of 40, which raised a red flag for suppliers. "The chip equipment manufacturers rated this at a much higher risk than we had," Stulen says.

VNL researchers identified what they called "seven deadly showstoppers," but by late 1998, at another industry session, solutions to many of these problems—such as how to make supersmooth mirrors—had been found, propelling EUV into first place when it came time to vote. "The group went from having an attitude of 'Sure, sure, tell us you can do that' to placing us up front," Freeman says.

Intel has also brought a get-the-job-done kind of urgency to laboratory employees unaccustomed to commercial deadlines. Peter J. Silverman, Intel's director of lithography capital equipment development, pushed forward by six to nine months the current circuit-printing demonstration and specified that the number of wafers produced by an EUV machine should be doubled. By moving the schedule, Intel has attempted to rally the industry around EUV and to eliminate electron projection lithography (EPL). "We fervently believe that there are not enough resources in the industry to develop both technologies," he says.

Silverman is also ready to blast ahead by placing an order with ASML for a $30-million EUV prototype machine, forcing the equipment manufacturer to commit to a delivery schedule.

It behooves Intel to push. Although AMD, Motorola, Infineon and Micron are partners, Intel negotiated contract terms that let it get the first machines produced and, because it is the largest investor in the $250-million program, the greatest number of tools.

Suppliers have to implement fully two crushingly difficult generations of technology before they finish making an investment of perhaps $750 million to start producing EUV machines. Getting them to buy into the breakneck schedule set by Intel may be a bigger challenge than creating angstrom-smooth mirrors. Even ASML, which dropped its involvement with EPL, is cautious, saying existing optical technologies may last longer than the industry expects. "It's too early to decide whether EUV will happen in the time frame Intel is pushing," says Jos Benschop, research manager at ASML.

Intel would also like to bring Nikon, its other main supplier, into the fold. But the industry's largest equipment manufacturer, which is researching EUV outside of the U.S. consortium, is not ready to commit to a single technology—and it continues work on EPL with IBM. Other chipmakers, such as Motorola and Texas Instruments, have voiced support for the EUV competitor. "It's still a horse race between EPL and EUV," says Gilbert L. Varnell, president and chief operating officer of Nikon Research Corporation of America. "Intel has taken the position that there's only one technology and they want to get rid of the competition. I'm not convinced that's the best approach for the industry. What if [EUV] fails? We're a toolmaker and they're a chipmaker, and there's a lot of other things we have to consider, such as manufacturability of the lithography equipment and profitability." Adds Lloyd R. Harriott, a former Bell Labs employee who headed the EPL program and worked on the early EUV program: "I think a lot of progress has been made with EUV. But they've got a really long way to go. There's a lot of marketing hype about how this is a done deal."

Varnell also believes that the current schedule—making commercial chips with EUV in 2005—is unrealistic, citing the nine years it took Nikon to develop the laser used in the current generation of lithography, a much less ambitious project. Says Varnell: "You're going from an image to full-up production system by 2005, and it is going to come from the national labs. I've been around the toolmaking business for a long time. I don't believe that's going to happen."

Along the way, another hurdle Intel and company have faced is convincing Washington to let a foreign company, the Dutch supplier ASML, enter the consortium. Four years ago the only major American tool supplier in the consortium was SVG. Ultratech Stepper, an early U.S. partner in EUV research, had to settle grudgingly for a minor role when it was viewed as lacking the necessary financial resources to develop an EUV product line. ASML, moreover, has subsequently bought SVG, which would leave ASML as the primary beneficiary of this technology transfer. Intel has "done everything in their power to give the technology on a silver platter to ASML," says David A. Markle, chief technology officer of Ultratech Stepper, adding that "Intel has approached this situation with the attitude that what's good for Intel is good for America."

Despite the trail of bruised egos, the EUV experience may serve as a case study for future research. It is one of the most successful collaborations between industry and national laboratories. More broadly, it constitutes a model for the creation of virtual laboratories that can undertake major projects on an as-needed basis without the huge overhead of a central research facility.

Whether Intel's buy-it-when-you-need-it strategy can work more generally remains to be seen. The real test may come in 15 years or so if EUV or EPL gives out and some wholly new substitute for silicon chips is needed. A paradigm shift—using molecules of DNA, nanotubes, quantum dots or other exotic materials to execute computations—may determine whether the virtual-research model can succeed. "Intel did a magnificent job of picking up the technology, recognizing its worthiness and driving it home," Freeman says. "But they're not putting the same effort into asking the questions about what to do when you get to 100 angstroms [10 nanometers]." Maybe one of Moore's successors will have to lay down the law for quantum computing.

The ever-shrinking size of microprocessors puts its transistors under the stress of increasingly powerful electrical fields. The next-generation microchips that may be at the heart of future supercomputers may replace traditional transistor design with an array of new innovations.

The Future of the Transistor

Robert W. Keyes

I am writing this article on a computer that contains some 10 million transistors, an astounding number of manufactured items for one person to own. Yet they cost less than the hard disk, the keyboard, the display and the cabinet. Ten million staples, in contrast, would cost about as much as the entire computer. Transistors have become this cheap because during the past 40 years engineers have learned to etch ever more of them on a single wafer of silicon. The cost of a given manufacturing step can thus be spread over a growing number of units.

How much longer can this trend continue? Scholars and industry experts have declared many times in the past that some physical limit exists beyond which miniaturization could not go. An equal number of times they have been confounded by the facts. No such limit can be discerned in the quantity of transistors that can be fabricated on silicon, which has proceeded through eight orders of magnitude in the years since the transistor was invented in 1948.

I do not have a definitive answer to the question of limits. I do, however, have some thoughts on how the future of solid-state electronics will develop and what science is needed to support continuing progress.

Several kinds of physical limitations might emerge as the size of the transistor continues to shrink. The task of connecting minute elements to one another might, for example, become impossible. Declining circuit size also means that researchers must cope with ever stronger electrical fields, which can affect the movement of electrons in many ways. In the not too distant future the transistor may span only hundreds of angstroms. At that point, the presence or absence of single atoms, as well as their behavior, will become significant. Diminishing size leads to increasing density of transistors on a chip, which raises the amount of waste heat thrown off. As the size of circuit elements drops below the wavelength of usable forms of radiation, existing manufacturing methods may reach their limits.

To see how such problems might arise and how they can be addressed, it is useful to review the operation of the field-effect transistor, the workhorse of modern data processing. Digital computers operate by manipulating statements made in a binary code, which consists of ones and zeroes. A field-effect transistor is operated so that, like a relay, it is switched only "on" or "off." The device therefore represents exactly one binary unit of information: a bit. In a large system, input signals control transistors that switch signal voltages onto output wires. The wires carry the signals to other switches that produce outputs, which are again sent on to another stage. The connections within the computer determine its function. They control the way that the inputs are transformed to become outputs, such as a word in a document or an entry in a spreadsheet.

The field-effect transistor contains a channel that interacts with three electrodes: a source, which supplies electrons to the channel; a drain, which receives them at the other side; and a gate, which influences the conductivity of the channel. Each part contains different impurity atoms, or dopants, which modify the electrical properties of the silicon.

The gate switches the transistor on when a positive voltage applied to it attracts electrons to the interface between the semiconductor and the gate insulator. These electrons then establish a connection between the source and drain electrodes that allows current to be passed between them. At this point, the transistor is "on." The connection persists for as long as the positive charge remains on the gate. An incoming signal is applied to the gate and thus determines whether the connection between source and drain is established. If a connection results, the output is connected to the ground potential, one of the standard digital voltages. If no connection results, the output is connected through the resistor to the positive power supply, the other standard digital voltage.

Circuits of transistors must be oblivious to the operations of neighboring arrays. Existing concepts of insulation, impedance and other basic electrical properties of semiconductors and their connections should work well enough, for designers' purposes, in the next generation of devices. It is only when conducting areas approach to within about 100 angstroms of one another that quantum effects, such as electron tunneling, threaten to create problems. In laboratory settings, researchers are already at the brink of this limit, at about 30 angstroms; in commercial devices, perhaps a decade remains before that limit is reached.

Another challenge is the strengthening of the electrical field that inevitably accompanies miniaturization. This tendency constrains the design of semiconductor devices by setting up a basic conflict. Fields must continually get stronger as electron pathways shrink, yet voltages must remain above the minimum needed to overwhelm the thermal energy of electrons. In silicon at normal operating temperatures, the thermal voltage is 0.026 electron volt. Therefore, whenever a semiconductor is switched so as to prevent the passage of electrons, its electrical barrier must be changed by a factor several times as large. One can minimize the thermal problem by chilling the chip (which becomes an expensive proposition).

Even cooling cannot end the problem of the electrical field. Signals must still have the minimum voltage that is characteristic of a semiconductor junction. In silicon this electrical barrier ranges between half a volt and a volt, depending on the degree of doping. That small voltage, applied over a very short distance, suffices to create an immensely strong electrical field. As electrons move through such a field, they may gain so much energy that they stimulate the creation of electron-hole pairs, which are themselves accelerated. The resulting chain reaction can trigger an avalanche of rising current, thereby disrupting the circuit. Today's chips push the limits in the quest for high speed, and electrical fields are usually close to those that can cause such avalanches.

Workers resort to a variety of tricks to mitigate the effects of strong electrical fields. They have designed field-effect transistors, for example, in which the field can be moved to a place where it does not disrupt other electronic functions. This stratagem is just one of many, all of which entail trade-offs with other desired characteristics, such as simplicity of design, ease of manufacture, reliability and long working life.

Miniaturization also increases the heat given off by each square centimeter of silicon. The reason is purely geometric: electrical pathways, and their associated energy losses, shrink in one dimension, whereas chip area shrinks in two. That relation means that as circuits get smaller, unit heat generation falls, albeit more slowly than does the number of units per square centimeter.

Devices already pour out as much as 30 watts per square centimeter, a radiance that one would expect of a material heated to about 1,200 degrees Celsius (this radiance value is about 10 times that of a range-top cooking surface in the home). Of course, the chips cannot be allowed to reach such

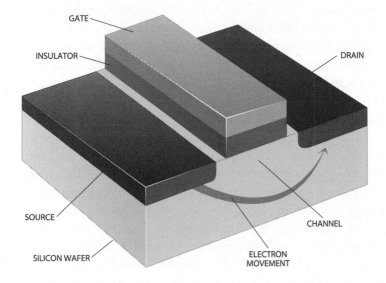

A field effect transistor contains a channel, a source, a drain and an insulated gate. When positive voltage is applied to the gate, electrons move near the insulation, establishing a connection underneath it, switching the transistor on.

temperatures, and so cooling systems remove heat as fast as it is produced. A variety of cooling technologies have been devised, including some rather intense ones. But the cost of using them in transistor circuits increases rapidly when the density of heat increases.

The exigencies of manufacturing impose constraints on the performance of electronic devices that might not be apparent from a purely theoretical discussion. Low-cost manufacturing results in small differences among the devices that are made on each wafer, as well as among those that are fabricated on different wafers. This variability cannot be banished—it is inherent in the way solid-state devices are made.

A semiconducting material, such as silicon, is made into a transistor in an integrated process involving many steps. Templates, called masks, are applied to the silicon in order to expose desired areas. Next, various operations involving chemical diffusion, radiation, doping, sputtering or the deposition of metal act on these areas, sometimes by constructing device features, other times by erecting scaffolding to be used in succeeding steps and then torn down. Meanwhile other devices—resistors, capacitors and conductors—are being built to connect the transistors.

Variations intrude at every step. For example, perfect focusing of the source of radiation over a large wafer is hard to achieve. The temperature of the wafer may vary slightly from one place to another during processing steps, causing a difference in the rate of chemical reactions. The mixing of gases in a reaction chamber may not be perfect. For many reasons, the properties of devices on a given wafer and between those on different wafers are not identical. Indeed, some devices on a wafer may be no good at all; the proportion of such irremediable errors places a practical limit on the size of an integrated circuit.

A certain amount of fuzziness is inherent in optical exposures. The light used in photolithography is diffracted as it passes through the holes in the template. Such diffraction can be minimized by resorting to shorter wavelengths.

When photolithographic fabrication was invented in the early 1970s, white light was used. Workers later switched to monochromatic laser light, moving up the spectrum until, in the mid-1980s, they reached the ultraviolet wavelengths. Now the most advanced commercial chips are etched by deep ultraviolet light, a difficult operation because it is hard to devise lasers with output in that range. The next generation of devices may require x-rays. Indeed, each generation of circuitry requires manufacturing equipment of unprecedented expense.

Other problems also contribute to the cost of making a chip. The mechanical controls that position wafers must become more precise. The "clean rooms" and chambers must become ever cleaner to ward off the ever smaller motes that can destroy a circuit. Quality-control procedures must become even more elaborate as the number of possible defects on a chip increases.

Miniaturization may at first glance appear to involve manipulating just the width and breadth of a device, but depth matters as well. Sometimes the third dimension can be a valuable resource, as when engineers sink capacitors edgewise into a chip to conserve space on the surface. At other times, the third dimension can constrain design. Chip designers must worry about the aspect ratio—that is, the relation of depth to surface area. The devices and connections on chips are built up in the silicon and on the surface as a series of layers resembling a sandwich. Part of the art of making devices smaller comes from using more layers. But the more layers there are, the more carefully controlled each must be, because each is affected by what is beneath it. The number of layers is limited by the costs of better control and more connections between layers.

The formulas that are used to design large devices cannot be used for the tiny transistors now being made in laboratories. Designers need to account for exotic new phenomena that appear in such extremely small devices. Because the effects cannot be accurately treated by purely analytic methods, the designers must have recourse to computer models that are able to simulate the motion of electrons in a device.

A computer follows a single electron through a device, keeping track of its position as time is increased in small steps. Physical theory and experimental information are used to calculate the probability of the various events that are possible. The computer uses a table for the probabilities, stored in its memory, and a random number generator to simulate the occurrence of these events. For example, an electron is accelerated by an electrical field, and the direction of its motion might be changed by a collision with an impurity. Adding the results of thousands of electrons modeled in this fashion gives a picture of the response of the device.

Consider the seemingly trivial question of how to represent the motion of an electron within an electrical field. When path lengths were comparatively long, an electron quickly accelerated to the point at which collisions robbed it of energy as fast as the field supplied new energy. The particle therefore spent most of its time at a constant velocity, which can be modeled by a simple, linear equation. When path lengths became shorter, the electron no longer had time to reach a stable velocity. The particles now accelerate all the time, and the equations must account for that complication.

If such difficulties can arise in modeling a well-understood phenomenon, what lies ahead as designers probe the murky physics of the ultrasmall? Simulations can be no better than the models that physicists make of events that happen in small spaces during the short periods. To refine these models, researchers need to carry out experiments on femtosecond timescales.

Expanded knowledge of solid-state physics is required, because as chips grow more complex they require more fabrication steps, and each step can influence the next. For instance, when doping atoms are introduced into a crystal, they tend to attract, repel or otherwise affect the motion of other dopants. Such effects of dopants on other dopants are not well understood; further experiments and theoretical investigations are therefore needed. Chemical reactions that take place on the surface of a silicon crystal demand a supply of silicon atoms, a kind of fluid flow within the solid lattice; does such motion carry other constituents along with it? These questions did not concern designers of earlier generations of chips, because existing transistors were then large enough to swamp such ultramicroscopic tendencies.

The prospect of roadblocks aside, the transistor has only itself to blame for speculation about alternative technologies. Its extraordinary success in the 1950s stimulated an explosive

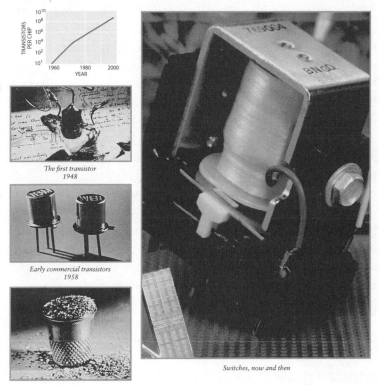

The first transistor
1948

Early commercial transistors
1958

Salt-size transistors
1964

Switches, now and then

Above, right: an electromagnetic switch, circa 1957. Transistors have become increasingly smaller over time and have allowed computer chips to manifest in smaller and smaller sizes.

development of solid-state physics. In the course of the work, investigators discovered many other phenomena, which in turn suggested a host of ideas for electronic devices. Several of these lines of research produced respectable bodies of new engineering knowledge but none that led to anything capable of even finding a niche in information processing.

Some workers have argued that the transistor owes its pre-eminence to having been the first off the block. Because of that head start, semiconductors have been the center of research, a position that guarantees them a margin of technological superiority that no rival can match. Yet I believe the transistor has intrinsic virtues that, in and of themselves, could probably preserve its dominant role for years to come.

I participated, as a minor player, in some of the efforts to build alternative switches, the repeated failures of which made me wonder what was missing. Of course, quite a few new fabrication methods had to be developed to implement a novel device concept. But even though these could be mastered, it was difficult to get a large collection of components to work together.

What gave the transistor its initial, sudden success? One difference stood out: the transistor, like the vacuum tube before it, has large gain. That is, it is capable of vastly amplifying signals of the kind processed in existing circuits, so that a small variation in input can produce a large variation in output. Gain makes it possible to preserve the integrity of a signal as it passes through many switches.

Rivals to the transistor may have been equally easy to miniaturize, but they exhibited far less gain. Take, for instance, bistable devices, which perform logic functions by moving between two stable states that are separated by an unstable transition. Researchers have produced such a transition by designing circuits having a range of values in which current declines as voltage increases. Any slight disturbance, such as that obtained by injecting extra current through the device, will switch the circuit between its two stable states.

Because this slight input can bring about large changes in the current and voltages, there is a sense in which gain is achieved. Yet the gain is far less useful than that provided by an ordinary transistor because it operates within rather narrow tolerances. A bistable switch thus performs deceptively well in the laboratory, where it is possible to fine-tune the circuit so it stays near enough to the crossover point. A collection of such switches, however, does not lend itself to such painstaking adjustments. Because not all the circuits will work, no complex device can be based on their operation. Negative resistance therefore plays no role in practical data processing.

The same difficulty has plagued the development of nonlinear optical devices, in which the intensity of optical beams replaces the currents and voltages of electrical circuits. Here, too, the operation depends on fine-tuning the system so that a small input will upset a delicate balance. (Such switches have occasionally been termed optical transistors, a label that misconstrues the principles of transistor action.)

Optical switches face a problem even more fundamental. Light, unlike electricity, hardly interacts with light, yet the interaction of signals is essential for logic functions. Optical signals must therefore be converted into electrical ones in a semiconductor. The voltage thus produced changes the optical response of another material, thereby modulating a beam of light.

Useful Interference

Another proposed switch, sometimes called a quantum interference device, depends on the interference of waves. In the most familiar case, that of electromagnetic radiation, or light, one wave is divided into two components. The components begin oscillating in phase—that is, their peaks and troughs vibrate in tandem. If the components follow routes of different lengths before reuniting, the phase relation between their waveforms will be changed. Consequently, the peaks and troughs either cancel or reinforce one another, producing a pattern of bright and dark fringes. The displacement of the fringes measures the relative phase of the system.

Electrons also possess a wave nature and can be made to interfere. If the two components of a wave move at equal speeds over similar paths to a rendezvous, they will reconstitute the original wave; if they move at different speeds, they will interfere. One can manipulate the velocity of one wave by applying a tiny electrical field to its pathway. The correct field strength will cause the waves to cancel so that no current can flow through the device.

At first sight, this action duplicates a field-effect transistor, which uses an electrical field to control a current through a semiconductor. In an interference device, however, conditions must be just right: if the applied voltage is too high or too low, there will be some current. This sensitivity means that an interference device will not restore the binary nature of a degraded input signal but will add its own measure of noise. Data passing from one such device to another will quickly degenerate into nothingness.

The Only Game in Town

The lack of real rivals means that the future of digital electronics must be sought in the transistor. The search begins anew with each voyage into a smaller scale or a different material. The latest reexamination was occasioned by the introduction of new semiconductor materials, such as gallium arsenide and related compounds, several of which may even be incorporated to achieve some desired characteristic in a single device. These combinations may be used to produce what are called heterojunctions, in which crystalline lattices of different energy gaps meet. Lattices may mesh imperfectly, creating atomic-scale defects, or they may stretch to one another, creating an elastic strain. Either defects or strain can produce electrical side effects.

These combinations complicate the physics but at the same time provide a variable that may be useful in surmounting the many design problems that miniaturization creates. For instance, the dopants that supply electrons to a semiconductor also slow the electrons. To reduce this slowing effect, one can alternate layers of two semiconductors in which electrons have differing energies. The dopants are placed in the high-energy semiconductor, but the electrons they donate immediately fall into the lower-energy layers, far from the impurities.

What, one may ask, would one want with a technology that can etch a million transistors into a grain of sand or put a supercomputer in a shirt pocket? The answer goes beyond computational power to the things such power can buy in the emerging information economy. It has only recently been taken for granted that anyone with a personal computer and a modem can search 1,000 newspapers for references to anything that comes to mind, from kiwifruit to quantum physics. Will it soon be possible for every person to carry a copy of the Library of Congress, to model the weather, to weigh alternative business strategies or to checkmate Garry Kasparov?

Predicting the future is a notoriously tricky undertaking. In his essay, David A. Patterson projects a future where the silicon chip is still the processor of choice, enhanced by using strategies such as parallel processing, pipelining, and superscalar approaches.

Microprocessors in 2020

David A. Patterson

U nlike many other technologies that fed our imaginations and then faded away, the computer has transformed our society. There can be little doubt that it will continue to do so for many decades to come. The engine driving this ongoing revolution is the microprocessor, the sliver of silicon that has led to countless inventions, such as portable computers and fax machines, and has added intelligence to modern automobiles and wristwatches. Astonishingly, the performance of microprocessors has improved 25,000 times over since their invention only 32 years ago.

I have been asked to describe the microprocessor of 2020. Such predictions in my opinion tend to overstate the worth of radical, new computing technologies. Hence, I boldly predict that changes will be evolutionary in nature, and not revolutionary. Even so, if the microprocessor continues to improve at its current rate, I cannot help but suggest that 25 years from now these chips will empower revolutionary software to compute wonderful things.

* * *

Two inventions sparked the computer revolution. The first was the so-called stored program concept. Every computer system since the late 1940s has adhered to this model, which prescribes a processor for crunching numbers and a memory for storing both data and programs. The advantage in such a system is that, because stored programs can be easily interchanged, the same hardware can perform a variety of tasks. Had computers not been given this flexibility, it is probable that they would not have met with such widespread use. Also, during the late 1940s, researchers invented the transistor. These silicon switches were much smaller than the vacuum tubes used in early circuitry. As such, they enabled workers to create smaller—and faster—electronics.

More than a decade passed before the stored program design and transistors were brought together in the same machine, and it was not until 1971 that the most significant pairing—the Intel 4004—came about. This processor was the first to be built on a single silicon chip, which was no larger than a child's fingernail. Because of its tiny size, it was dubbed a microprocessor. And because it was a single chip, the Intel 4004 was the first processor that could be made inexpensively in bulk.

The method manufacturers have used to mass-produce microprocessors since then is much like baking a pizza: the dough, in this case silicon, starts thin and round. Chemical toppings are added, and the assembly goes into an oven. Heat transforms the toppings into transistors, conductors and insulators. Not surprisingly, the process—which is repeated perhaps 20 times—is considerably more demanding than baking a pizza. One dust particle can damage the tiny transistors. So, too, vibrations from a passing truck can throw the ingredients out of alignment, ruining the end product. But provided that does not happen, the resulting wafer is divided into individual pieces, called chips, and served to customers.

Although this basic recipe is still followed, the production line has made ever cheaper, faster chips over time by churning out larger wafers and smaller transistors. This trend reveals an important principle of microprocessor economics: the more chips made per wafer, the less expensive they are. Larger chips are faster than smaller ones because they can hold more transistors. The Intel Pentium II, for example, contains 7.5 million transistors and is much larger than the Intel 4004, which had a mere 2,300 transistors. But larger chips are also more likely to contain flaws. Balancing cost and performance, then, is a significant part of the art of chip design.

Most recently, microprocessors have become more powerful, thanks to a change in the design approach. Following the lead of researchers at universities and laboratories across the U.S., commercial chip designers now take a quantitative approach to computer architecture. Careful experiments precede hardware development, and engineers use sensible metrics to judge their success. Computer companies acted in concert to adopt this design strategy during the 1980s, and as a result, the rate of improvement in microprocessor technology has risen from 35 percent a year only a decade ago to its current high of approximately 55 percent a year, or almost 4 percent each month. Processors are now four times faster than had been predicted in the early 1980s.

In addition to progress made on the production line and in silicon technology, microprocessors have benefited from recent gains on the drawing board. These breakthroughs will undoubtedly lead to further advancements in the near future. One key technique is called pipelining. Anyone who has done laundry has intuitively used this tactic. The nonpipelined approach is as follows: place a load of clothes in the washer. When the washer is done, place the load into the dryer. When the dryer is finished, fold the clothes. After the clothes are put away, start all over again. If it takes an hour to do one load this way, 20 loads take 20 hours.

The pipelined approach is much quicker. As soon as the first load is in the dryer, the second dirty load goes into the washer, and so on. All the stages operate concurrently. The pipelining paradox is that it takes the same amount of time to clean a single dirty sock by either method. Yet pipelining is faster in that more loads are finished per hour. In fact, assuming that each stage takes the same amount of time, the time saved by pipelining is proportional to the number of stages involved. In our example, pipelined laundry has four stages, so it would be nearly four times faster than nonpipelined laundry. Twenty loads would take roughly five hours.

Similarly, pipelining makes for much faster microprocessors. Chip designers pipeline the instructions, or low-level commands, given to the hardware. The first pipelined microprocessors used a five-stage pipeline. (The number of stages completed each second is given by the so-called clock rate. A personal computer with a 200-megahertz clock then executes 200 million stages per second.) Because the speedup from pipelining equals the number of stages, recent microprocessors have adopted eight or more stage pipelines. One 1997 microprocessor uses this deeper pipeline to achieve a 600-megahertz clock rate. As machines head toward the next century, we can expect pipelines having even more stages and higher clock rates.

Also in the interest of making faster chips, designers have begun to include more hardware to process more tasks at each stage of a pipeline. The buzzword "superscalar" is commonly used to describe this approach. A superscalar laundromat, for example, would use a professional machine that could, say, wash three loads at once. Modern superscalar microprocessors try to perform anywhere from three to six instructions in each stage. Hence, a 250-megahertz, four-way superscalar microprocessor can execute a billion instructions per second. A 21st-century microprocessor may well launch up to dozens of instructions in each stage.

Despite such potential, improvements in processing chips are ineffectual unless they are matched by similar gains in memory chips. Since random-access memory (RAM) on a chip became widely available in the mid-1970s, its capacity has grown fourfold every three years. But memory speed has not increased at anywhere near this rate. The gap between the top speed of processors and the top speed of memories is widening.

One popular aid is to place a small memory, called a cache, right on the microprocessor itself. The cache holds those segments of a program that are most frequently used and thereby allows the processor to avoid calling on external memory chips much of the time. Some chips actually dedicate more transistors to the cache than they do to the processor itself. Future microprocessors will allot most resources to the cache to better bridge the speed gap.

The Holy Grail of computer design is an approach called parallel processing, which delivers the benefits of a single fast processor by engaging many inexpensive ones at the same time. In our analogy, we would go to a laundromat and use 20 washers and 20 dryers to do 20 loads simultaneously. Clearly, parallel processing is a costly solution for small workloads. And writing a program that can use 20 processors at once is much harder than distributing laundry to 20 washers. Indeed, the program must specify which instructions can be launched by which processor at what time.

Superscalar processing bears similarities to parallel processing, and it is more popular because the hardware automatically finds instructions that launch at the same time. But its potential processing power is not as large. If it were not so difficult to write the necessary programs, parallel processors could be made as powerful as one could afford. For the past 25 years, computer scientists have predicted that the programming problems will be overcome.

In reviewing old articles, I have seen fantastic predictions of what computers would be like in 1997. Many stated that optics would replace electronics; computers would be built entirely from biological materials; the stored program concept would be discarded. These descriptions demonstrate that it is impossible to foresee what inventions will prove commercially viable and go on to revolutionize the computer industry. In my career, only three new technologies have prevailed: microprocessors, random-access memory and optical fibers. And their impact has yet to wane, decades after their debut.

Surely one or two more inventions will revise computing in the next 25 years. My guess, though, is that the stored program concept is too elegant to be easily replaced. I believe future computers will be much like machines of the past, even if they are made of very different stuff. I do not think the microprocessor of 2020 will be startling to people from our time, although the fastest chips may be much larger than the very first wafer, and the cheapest chips may be much smaller than the original Intel 4004.

Pipelining, superscalar organization and caches will continue to play major roles in the advancement of microprocessor technology, and if hopes are realized, parallel processing will join them. What will be startling is that microprocessors will probably exist in everything from light switches to pieces of paper. And the range of applications these extraordinary devices will support, from voice recognition to virtual reality, will very likely be astounding.

Today microprocessors and memories are made on distinct manufacturing lines, but it need not be so. Perhaps in the near future, processors and memory will be merged onto a single chip, just as the microprocessor first merged the separate components of a processor onto a single chip. To narrow the processor-memory performance gap, to take advantage of parallel processing, to amortize the costs of the line and simply to make full use of the phenomenal number of transistors that can be placed on a single chip, I predict that the high-end microprocessor of 2020 will be an entire computer.

Let's call it an IRAM, standing for intelligent random-access memory, since most of the transistors on this merged chip will be devoted to memory. Whereas current microprocessors rely on hundreds of wires to connect to external memory chips, IRAMs will need no more than computer network connections and a power plug. All input-output devices will be linked to them via networks. If they need more memory, they will get more processing power and network connections as well, and vice versa—an arrangement that will keep the memory capacity and processor speed and network connectivity in balance. IRAMs are also the ideal building block for parallel processing. And because they would require so few external connections, these chips could be extraordinarily small. We may well see cheap "picoprocessors" that are smaller than the ancient Intel 4004. If parallel processing succeeds, this sea of transistors could also be used by multiple processors on a single chip, giving us a micromultiprocessor.

Today's microprocessors are more than 100,000 times faster than their 1950s ancestors, and when inflation is considered, they cost 1,000 times less. These extraordinary facts explain why computing plays such a large role in our world now. Looking ahead, microprocessor performance will easily keep doubling every 18 months through the turn of the century. After that, it is hard to bet against a curve that has outstripped all expectations. But it is plausible that we will see improvements in the next 25 years at least as large as those seen in the past 50. This estimate means that one desktop computer in 2020 will be as powerful as all the computers in Silicon Valley today. Polishing my crystal ball to look yet another 25 years ahead, I see another quantum jump in computing power.

The implications of such a breathtaking advance are limited only by our imaginations. Fortunately, the editors have asked others to ponder the possibilities, and I happily pass the baton on to them.

Intimidating technical obstacles wait for silicon-based microprocessors as the pressure for faster and smaller chips increases. One possible way to extend the usefulness of silicon chips would be to enhance their mostly two-dimensional architecture into multi-layered three-dimensional space.

A Vertical Leap for Microchips

Thomas H. Lee

The city of San Francisco stretches over 45 square miles—about twice the area of the island of Manhattan. Yet the economic output of Manhattan dwarfs that of San Francisco. A principal reason for the disparity is that offices in earthquake-prone California tend to spread their workers and machines close to ground level, whereas businesses in New York are stacked vertically into the skies. By building upward rather than outward, developers increase not only the value of their real estate but also the working power of the city as a whole.

An analogous strategy applied to the microscopic world of computer chips could rejuvenate a semiconductor industry that has recently begun to show signs of senescence. Surprisingly, of the more than 100 quadrillion transistors that Intel co-founder Gordon E. Moore estimates have been produced to date, nearly every one has been built on the "ground level," directly on the surface of silicon crystals. Engineers have accomplished a fantastically regular doubling of transistor density per microchip—we call it Moore's Law in the industry—

simply by expanding the area of each chip and shrinking the size of each transistor. This is like building only shopping malls and no skyscrapers.

That is about to change. For one thing, physicists tell us that Moore's Law will end when the gates that control the flow of information inside a chip become as small as the wavelength of an electron (on the order of 10 nanometers in silicon), because transistors then cease to "transist." And many intimidating technical obstacles loom between the current state of the art and that fundamental limit. The trajectory of progress has already begun to droop.

Fortunately, I and other engineers have recently found a way to skirt some of those obstacles, to give Moore's Law a new breath of life and even to accelerate the delivery of more computing power for less cost. We have shown that it is feasible to make chips that contain vertical microcircuits using the same semiconductor foundries, the same standard materials and similar techniques to those used to manufacture conventional computer chips.

Such "three-dimensional" chips are now being commercialized by Matrix Semiconductor, a company I co-founded in 1998 in Santa Clara, Calif., with computer scientist P. Michael Farmwald and chip design expert Mark C. Johnson. Sometime in 2002, 3-D memory circuits will hit the market. They will be just the first of a new generation of dense, inexpensive chips that promise to make digital recording media both cheap and convenient enough to replace photographic film and audiotape. In laboratories at Stanford University and Matrix, we have also created prototype devices that incorporate vertical logic circuits. There seems to be good reason to expect that even for microprocessors, the sky is the limit.

Today's state-of-the-art microcircuits are not entirely two-dimensional. Intel's Pentium 4 processor, for example, boasts seven layers of wiring, embedded within patterns of insulating material. It is only on the bottom layer of pure silicon, however, that the active semiconducting regions lie.

So far the industry has managed to sustain Moore's Law largely by improving the way it uses that silicon wafer. Materials scientists have invented ways to grow giant crystals of silicon 30 centimeters in diameter that contain less than one part per billion of impurities. Clean-room robots shoot carefully metered doses of ions into wafers cut from these crystals. A process called photolithography defines the ion-activated regions with patterns of light and acid etching to make transistors. To cram more transistors onto one wafer requires light of ever shorter wavelength. Mercury vapor lamps have been replaced by deep-ultraviolet excimer lasers that inscribe 130-nanometer features and can put a billion transistors on a chip. Further improvement should push that limit to 65 nanometers and perhaps 16 billion transistors.

The road beyond that point may be rough, however. Extreme ultraviolet lithography systems that use even shorter wavelengths are just now beginning to function in the laboratory. They still pose many significant problems [as noted in "Getting More from Moore's," page 19].

If history is any guide, engineers will probably clear these hurdles; the economic incentive to do so is huge. But as the number of obstacles increases, the pace of progress may slow considerably. The official "road map" published by the Semiconductor Industry Association projects that chips will grow 4 to 5 percent a year in area; historically, area has grown about 15 percent a year. The periodic 30 percent reduction in minimum feature size is probably now going to occur every three years instead of every two. Even at this slower pace, Moore's Law will most likely hit fundamental limits sometime between 2010 and 2020.

One important factor has remained roughly constant: the cost of semiconductor real estate, at about $1 billion per acre of processed silicon. So why haven't silicon developers taken the seemingly obvious step of building upward? The simplest reason is that transistors are fastest and most reliable only when formed from the perfectly aligned atoms of a wafer cut from a single crystal of silicon.

Once we coat that semiconducting wafer with an insulating oxide or metal wires, there is no known way to recover the underlying crystalline pattern—it's like trying to match the pattern of a parquet floor after it has been covered with carpet. Silicon deposited onto a noncrystalline surface tends to be completely disordered and amorphous. With appropriate heat treatment, we can encourage the silicon to form minuscule islands ("grains") of single crystals, but the ordered lines of atoms collide abruptly at odd angles at the boundaries between grains. Contaminants can pile up at these barriers and short out any transistor or memory cell caught in the middle. For many years, such amorphous and polysilicon (short for polycrystalline silicon) devices were so poor that no one seriously considered them for anything more sophisticated than solar cells.

In the early 1980s, however, premature worries that Moore's Law was about to fail stimulated a flurry of attempts to make 3-D microcircuits in which the transistors spanned vertical towers—rather than horizontal bridges—of silicon. James F. Gibbons and others at Stanford used laser beams to improve the quality of silicon films deposited onto nonsilicon substrates. Others tried stacking conventional 2-D chips on top of one another. Regrettably, the former approach was too slow and the latter was too expensive to be economically competitive. Traditional chipmaking stayed on track, and engineers stopped thinking much about vertical circuits.

In 1997 Farmwald and I started exploring 3-D chips again and realized that two key enabling technologies, developed for other purposes, made 3-D circuits truly practical for the first time. One was a technique to lay down polysilicon so that each island of a single crystal is large enough to encompass many

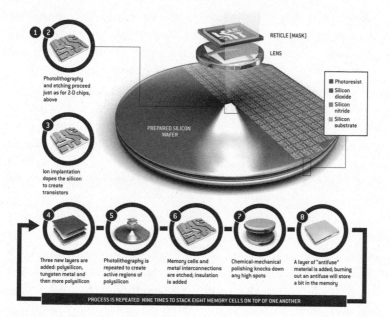

RETICLE (MASK)

LENS

Photolithography
and etching proceed
just as for 2-D chips,
above

Ion implantation
dopes the silicon
to create
transistors

PREPARED SILICON
WAFER

■ Photoresist
■ Silicon
   dioxide
■ Silicon
   nitride
■ Silicon
   substrate

Three new layers are
added: polysilicon,
tungsten metal and
then more polysilicon

Photolithography is
repeated to create
active regions of
polysilicon

Memory cells and
metal interconnections
are etched; insulation
is added

Chemical-mechanical
polishing knocks down
any high spots

A layer of "antifuse"
material is added; burning
out an antifuse will store
a bit in the memory

PROCESS IS REPEATED NINE TIMES TO STACK EIGHT MEMORY CELLS ON TOP OF ONE ANOTHER

Three-D chips can be manufactured using techniques perfected in today's 2-D chip design. Techniques such as this can lend silicon-based microchips added life and allow them to dodge many of the problems associated with the speed and size limitations of silicon.

memory cells or transistors. The second advance was a way to flatten each coat of new material so that the chips don't rise unevenly like towers built by drunken bricklayers.

We can thank the flat-panel-display industry for the first breakthrough. Its engineers figured out how to make millions of transistors from a thin film spread over a large, amorphous substrate (glass, in their case; other materials in ours). Thin-film transistors now populate the display panels of virtually every laptop computer. Part of the secret is to deposit the silicon at about 400 degrees Celsius as an extremely smooth (though amorphous) film, then to cook the entire wafer uniformly above about 500 degrees C for a few minutes. This converts the film to polysilicon with regular crystalline regions of a micron or more in diameter. Although LCD panels require only a single layer of transistors, the same machines that make the panels can also manufacture multilayer devices.

The second key enabling advance, called chemical-mechanical polishing (or CMP), emerged from IBM's research labs in the late 1980s. Back then, chip designers considered it risky to add two or three layers of metal on top of the silicon wafer because each new layer added hills and valleys that made it difficult to keep photolithographic patterns in focus.

To eliminate the bumps in each layer, process technologists adapted a trick that lens makers use to polish mirrors. The basic technique was used on all Intel 80486 processors: after each coating of silicon, metal or insulating oxide is added, the wafer is placed facedown on a pad. Spindles then rotate the pad and wafer in opposite directions while a slurry of abrasives and reactive alkaline chemicals passes in between. After mere minutes of polishing, the wafer is flat to within 50 nanometers, an ideal substrate for further processing. With advances in CMP machines, seven and eight layers of metal have become common in microchip designs; patience seems to be the main limiting factor in adding still more layers.

Building directly on these 2-D technologies, we have made 3-D circuits by coating standard silicon wafers with many successive layers of polysilicon (as well as insulating and metallic layers), polishing the surface flat after each step. Although electrons do not move quite as easily in polysilicon as they do in the single-crystal kind, research has produced 3-D transistors with 90 to 95 percent of the electron mobility seen in their 2-D counterparts.

Stacking devices vertically offers a way around some of the weighty obstacles that threaten to derail Moore's Law. As shopping-mall-style chips continue to sprawl outward, for example, it becomes increasingly hard to keep the photolithographic image in focus at the edges. And the relatively long wires that connect far-flung sections of conventional microprocessors cause delays that reduce performance and complicate design.

Ever shrinking circuits pose other problems. Transistors depend critically on a thin insulating layer below the control electrode. In the most advanced 2-D chips, this layer of silicon dioxide insulation measures just three nanometers—about two dozen atoms—in thickness. From transistor to transistor, that thickness must not vary by more than one or two atoms. The industry routinely meets this challenge, because it is much easier to grow superthin films than it is to etch supernarrow channels. But there may be no practical way to make these insulating layers much thinner, because current flow by quantum tunneling makes them progressively worse insulators. It's likely that some other material will soon have to replace silicon dioxide, but toolmakers have yet to agree on what that material will be.

There have been many novel chip designs proposed to address these problems. Most depend on replacing silicon altogether with various exotic materials, such as organic polymers, carbon fullerenes, copper compounds, ferroelectrics or magnetic alloys. But to abandon silicon is to squander an enormously valuable foundation of knowledge constructed over 50 years with some $100 billion worth of investment.

The 3-D electronic design process, in contrast, introduces no new atoms and leverages the huge industry investment in thin-film and CMP equipment. Because it is so expensive to produce and process ultrapure silicon ingots, the cost of silicon is largely proportional to the area (not the volume) consumed. So vertical electronics can reduce manufacturing costs 10-fold or more compared with traditional chips. And the density of 3-D devices should increase at least as fast as Moore's Law as we add more and more layers.

Digital Film and Beyond

Traditionally, semiconductor companies have worked the bugs out of new fabrication processes by making memory devices before attempting to mass-produce more complicated chips such as logic circuits. Memories are vast arrays of fundamentally simple cells, so there are fewer skills to master and fewer problems to solve.

That is the approach we at Matrix will take later this year as we introduce a 3-D memory chip in which the cells are stacked eight high. Unlike the RAM memories used in PCs, these chips use exceedingly simple memory cells that make them more like film, indelible once written. They are intended to be a low-cost medium for digital photography and audio. With 512 million memory cells, this first vertical microchip has enough capacity to store more than an hour of high-quality audio (through data compression) and a few hundred photographs (each comprising about one million pixels). The capacity will rise, and the unit cost will fall, over time. We have already proved that 12-cell-high devices are feasible, and 16-layer chips seem well within reach.

We have also demonstrated much more complex 3-D microcircuits in the laboratory, including static RAM, logic gates and even erasable EPROM memories. Although they are in very early stages of development, these basic building blocks are all that is needed to recast any planar circuit—including dynamic RAM, nonvolatile memories, wireless transceivers, and microprocessors—in 3-D form. Stood on end, the transistors in such circuits could be quite tiny because their channels will be made from thin films that are 10 times as precise as channels defined by ultraviolet light.

As with all engineering advances, this new manufacturing technique has limitations and trade-offs. Some fraction of memory cells or transistors in a vertical microcircuit will happen to straddle a boundary between polysilicon grains and will possibly fail as a result. We will have to use error detection and correction routines, like those used with audio CDs, and find ways to route signals around defective paths. The strategies of fault-tolerant computing, though well known, have generally not been built into microchips themselves. Such techniques are unnecessary and too cumbersome for application in most planar contexts, but the cost reductions afforded by 3-D processing fortuitously make the remedial technology economically feasible precisely when it becomes necessary.

Speed is another trade-off. Modern thin-film transistors typically perform at about half the speed of monocrystalline devices, although the difference is smaller when you compare entire circuits, because components packed in three dimensions need considerably shorter wires. Numerous researchers are investigating ways to close that gap further.

Beyond those special considerations, 3-D chips face essentially the same challenges as do conventional planar electronics—certain problems just appear sooner because of the effective acceleration of Moore's Law. Heat may be the most acute issue for dense 3-D devices because of their smaller surface area. The power density of a modern microprocessor already exceeds that of the burner on a typical stove. Ineffectiveness of current strategies for dissipating all that heat, such as reducing voltages or selectively activating only parts of a circuit, may limit the performance of dense 3-D circuits unless more advanced cooling technology is used. Fortunately, the newest microrefrigerators can now remove 200 watts per square millimeter while consuming only about one watt. Thermal limits are thus not yet fundamental impediments.

There is certainly lots of room for improvement. The fluid-cooled human brain, whose dimensions considerably exceed those of any 3-D circuit currently contemplated, dissipates a mere 25 watts; a 2.2-square-centimeter Pentium 4 microprocessor, in contrast, consumes about 80 watts. Although we cannot rule out the possibility that the inability to solve the heat problem may ultimately impose harsh limits on what 3-D circuits can do, history suggests that the strong economic incentives at play will eventually spark creative solutions.

Enabling Moore's Law to continue even a few years longer than it otherwise would have will have far-reaching consequences. For 30 years, chip manufacturers have striven constantly to print ever smaller structures within a single plane. It seems inevitable that in the future we will scale microcircuits vertically as well as horizontally. The technology is both possible and practical, and the benefits are far too compelling to ignore.

Future growth of microchips may be possible by replacing silicon with other materials. Non-silicon matrixes, including aluminum indium gallium phosphide and gallium arsenide, might underpin the newest and fastest microchips.

The Semiconducting Menagerie

Ivan Amato

J ust as the Stone Age, at technology's dawning, was actually an age of stone and bone and hide and wood and whatever else our ancestors found to be useful, this Silicon Age of ours is much more than its moniker suggests. Gallium arsenide, indium phosphide, aluminum indium gallium phosphide and mercury cadmium telluride are but a few of the major players in an ever expanding cast of semiconducting characters that have been finding roles in exotic or ultraefficient lasers and blazingly fast electronic circuits.

Those who think they have never seen or used a nonsilicon semiconductor are wrong. Remember those funky red-light-emitting diodes (LEDs) and watch and calculator displays that first started appearing in the late 1960s? They were based on such materials as gallium arsenide phosphide and gallium phosphide. And LEDs were just the beginning. Practically every compact-disc player contains a two-dollar, mass-produced semiconductor laser, the "stylus" that bounces off the CD's microcode of pits and spaces before shining onto semiconductor photodetectors for eventual conversion into sound. The heart of this laser is an exquisitely thin stripe of gallium arsenide sandwiched between slices of gallium aluminum arsenide, which is more electrically insulating. The infrared light emitted from this sandwich is produced when electrons and positively charged electron deficiencies (called holes) recombine, annihilating one another and releasing photons.

Telephony is another stronghold of nonsilicon semiconductors. In order to transmit effectively at low power, cellular telephones operate at frequencies at the edge of, or beyond, the capabilities of silicon circuitry. Thus, most cellular telephones have gallium arsenide circuitry, within which electrons move faster than they do inside silicon, enabling higher-frequency operation. Hardwired telephone systems, too, rely on nonsilicon semiconductors. Costly high-quality semiconductor lasers made of indium gallium arsenide phosphide, for example, send light-encoded data and voice signals down optical fibers.

What most of these applications have in common is that they are all related to the generation and detection of light. "The one thing that silicon cannot do is produce light," laments Harvey Serreze, operations manager of the optoelectronics division of Spire Corporation, a high-tech company based near Boston. Like the majority of those working on nonsilicon semiconductors, Serreze and his colleagues concentrate on the so-called III-V semiconductors, which are made by mixing and matching one or more elements from columns III and V of the periodic table of chemical elements. The usual suspects from column III are aluminum, gallium and indium; column V staples include nitrogen, phosphorus and arsenic. The lure of III-V compound semiconductors—and their even more exotic "II-VI" cousins—is that each has its own set of intrinsic and potentially useful electronic or optical traits.

Underlying these properties is the material's so-called band gap. This term refers to a kind of forbidden territory for electrons associated with the atoms of a crystal. Below the band gap is the valence band, a range of lower-energy electron states in which the electrons in the crystal remain closely associated with a specific atom in the crystal. Above the band gap is the conduction band, consisting of higher-energy states in which electrons are no longer associated with individual atoms and can flow relatively freely. Hence, the band gap is the amount of energy, measured in electron volts, that is required to cause electrons to leap from the valence band, over the band gap and into the conduction band.

Once in the conduction band, the electrons can carry current or fall back across the band gap to the valence band. When this latter event occurs, the electron typically falls into an empty "bonding" site, or electron deficiency, vacated by another electron. This phenomenon, known as recombining with a hole, causes a photon to be emitted. Exploitation of these recombinations is the basic principle behind almost all semiconductor lasers.

The key feature in all this activity is the band gap. The energy, and therefore the wavelength, of the emitted photons is a function of the band gap: the wider this gap, the greater the energy involved in falling back across it and the shorter the wavelength of the emitted photon. The band gap also determines whether and under what conditions the semiconductor can detect light and under what range of temperatures it will be able to serve as a reliable part of a transistor in, say, a computer chip.

The chemical composition of each semiconductor determines its intrinsic band gap, but researchers have been coming up with ever more clever ways of coercing these materials to do what they otherwise would not. By sprinkling different amounts of nitrogen atoms into gallium phosphide crystals, for example, materials researchers can alter the crystal's band gap so that it emits either red or green light. Such techniques have given rise to most of the materials—and colors—in the huge, worldwide industry for light-emitting diodes. It has also led to an enormous menagerie of one-of-a-kind materials and devices that make for respectable technical publications but rarely have the right stuff for lucrative ventures.

"People have gone through the periodic table and have come up with exotic [semiconducting] compounds, most of which are too tough to make," Spire's Serreze remarks. Among the pitfalls are too many microscopic defects, which put an end to light-emitting recombinations of electrons and holes, and too much strain within the crystals, which shortens the material's useful lifetime. To avoid the common disappointment of finding coveted properties in impractical materials, theoreticians and experimenters have fostered a golden age in a field known as band-gap engineering.

Band-gap engineers grow fabulously complex semiconductor crystals using such sophisticated, state-of-the-art techniques as chemical vapor deposition and molecular-beam epitaxy. Not uncommonly, these crystals consist of hundreds of layers, many no more than several atoms wide and each consisting of a chemical compound that the crystal grower essentially dials in. A key advantage of band-gap engineering is that it allows researchers to get a desired band gap by varying a crystal's layer-by-layer architecture, rather than by concocting some very esoteric compound that will most likely have undesirable properties that render it useless.

Because the layer-by-layer structure, not the specific chemical composition, sets the band gap, engineers are free to use relatively mundane and well-behaved materials, such as gallium arsenide, gallium aluminum arsenide and indium phosphide. With quantum-mechanical reasoning and calculations to guide them, these band-gap engineers design crystals whose interiors are essentially programmed to manipulate electrons passing through them.

Even as nonsilicon semiconductors stake their claim to more territory in technology's landscape, silicon rests secure in its kingdom. For the time being, at least, silicon remains the semiconductor of choice for applications such as computer chips, in which only electronic motion really matters and there is no emission of light. Silicon has proved so fantastic for microelectronic applications that solid-state scientists are often moved to speak of the element in unabashedly reverential terms. "Silicon is God's gift to the world," exclaims Wolfgang Choyke, a physicist at the University of Pittsburgh.

Not that researchers haven't tried and even come close to establishing a gallium arsenide beachhead in silicon's microprocessing stronghold. The legendary computer designer Seymour Cray used gallium arsenide chips in his last machine, the Cray-3 supercomputer. The machine never got off the ground commercially, partly because of the intricate and expensive cooling system needed to counter the heat produced by the chips. "That was such a huge undertaking that it could not find customers," explains Ira Deyhimy, vice president of product development at Vitesse Semiconductor Corporation, one of the largest suppliers of gallium arsenide chips.

At least one company did bring a gallium arsenide-based, high-performance computer to market. In the early 1990s, Convex Computer Corporation in Richardson, Tex. (now the Convex Division of Hewlett Packard), shipped more than 60 such machines, each of which sold for several million dollars. According to Steve Wallach, who leads the division, Convex's initial tack was to build their machines using several cooler-running, lower-power gallium arsenide chips, which were also made by Vitesse. The problem, however, was that Vitesse could not produce the chips with a high enough yield to meet Convex's needs (nor could anyone else, Wallach adds).

Deyhimy, however, notes that some supercomputer companies are moving to hybrid architectures that combine the computing power of superdense silicon chips with the swifter chip-to-chip data-shuttling capabilities of gallium arsenide. "Our most complex part in production has 1.4 million transistors," he says, pointing out that even if gallium arsenide's role in microprocessing remains minor compared with silicon's, the material will grow impressively in telecommunications, data communications and other burgeoning information industries that depend more on fast switching speeds and high data rates or frequencies than on processing power.

As far as gallium arsenide has come lately in the stratosphere of high-speed electronics for the communications world, it may very well have competition one day from its old nemesis, silicon. The Defense Advanced Research Projects Agency, for one, is funding several research groups that are combining silicon with germanium or with germanium and carbon atoms in various semiconductor alloys and quantum-well structures.

Whatever you call this technological age, semiconductors of many flavors will infiltrate ever more of its nooks and crannies. When it comes to predicting what might be possible, caution has been the hazardous path. Says Serreze, "The naysayers are on the side of the road," whereas the visionaries are speeding by.

The small size of components manufactured with nanotechnology opens up whole new doors in virtually all fields of science. The ability to put millions of transistors on a surface the size of a human hair could revolutionize the construction of computer chips.

Engineering Microscopic Machines

Kaigham J. Gabriel

The electronics industry relies on its ability to double the number of transistors on a microchip every 18 months, a trend that drives the dramatic revolution in electronics. Manufacturing millions of microscopic elements in an area no larger than a postage stamp has now begun to inspire technology that reaches beyond the field that produced the pocket telephone and the personal computer.

Using the materials and processes of microelectronics, researchers have fashioned microscopic beams, pits, gears, membranes and even motors that can be deployed to move atoms or to open and close valves that pump microliters of liquid. The size of these mechanical elements is measured in microns—a fraction of the width of a human hair. And like transistors, millions of them can be fabricated at one time.

In the next 50 years, this structural engineering of silicon may have as profound an impact on society as did the miniaturization of electronics in preceding decades. Electronic computing and memory circuits, as powerful as they are, do nothing more than switch electrons and route them on their way over tiny wires. Micromechanical devices will supply electronic systems with a much needed window to the physical world, allowing them to sense and control motion, light, sound, heat and other physical forces.

The coupling of mechanical and electronic systems will produce dramatic technical advances across diverse scientific and engineering disciplines. Thousands of beams with cross sections of less than a micron will move tiny electrical scanning heads that will read and write enough data to store a small library of information on an area the size of a microchip. Arrays of valves will release drug dosages into the bloodstream at precisely timed intervals. Inertial guidance systems on a chip will aid in locating the position of military combatants and direct munitions precisely to targets.

Microelectromechanical systems, or MEMS, is the name given to the practice of making and combining miniaturized mechanical and electronic components. MEMS devices are made using manufacturing processes that are similar, and in some cases identical, to those required for crafting electronic components.

One technique, called surface micromachining, parallels electronics fabrication so closely that it is essentially a series of steps added to the making of a microchip. Surface micromachining acquired its name because the small mechanical structures are "machined" onto the surface of a silicon disk, known as a wafer. The technique relies on photolithography as well as other staples of the electronic manufacturing process that deposit or etch away small amounts of material on the chip.

Photolithography creates a pattern on the surface of a wafer, marking off an area that is subsequently etched away to build up micromechanical structures such as a motor or a freestanding beam. Manufacturers start by patterning and etching a hole in a layer of silicon dioxide deposited on the wafer. A gaseous vapor reaction then deposits a layer of polycrystalline silicon, which coats both the hole and the remaining silicon dioxide material. The silicon deposited into the hole becomes the base of the beam, and the same material that overlays the silicon dioxide forms the suspended part of the beam structure. In the final step, the remaining silicon dioxide is etched away, leaving the polycrystalline silicon beam free and suspended above the surface of the wafer.

Such miniaturized structures exhibit useful mechanical properties. When stimulated with an electrical voltage, a beam with a small mass will vibrate more rapidly than a heavier device, making it a more sensitive detector of motion, pressure or even chemical properties. For instance, a beam could adsorb a certain chemical (adsorption occurs when thin layers of a molecule adhere to a surface). As more of the chemical is adsorbed, the weight of the beam changes, altering the frequency at which it would vibrate when electrically excited. This chemical sensor could therefore operate by detecting such changes in vibrational frequency. Another type of sensor that employs beams manufactured with surface micromachining functions on a slightly different principle. It changes the position of suspended parallel beams that make up an electrical capacitor—and thus alters the amount of stored electrical charge—when an automobile goes through the rapid deceleration of a crash. Analog Devices, a Massachusetts-based semiconductor concern, manufactures this acceleration sensor to trigger the release of an air bag. The company has sold more than half a million of these sensors to automobile makers over the past two years.

This air-bag sensor may one day be looked back on as the microelectromechanical equivalent of the early integrated electronics chips. The fabrication of beams and other elements of the motion sensor on the surface of a silicon microchip has made it possible to produce this device on a standard integrated-circuit fabrication line.

In microelectronics the ability to augment continually the

number of transistors that can be wired together has produced truly revolutionary developments: the microprocessors and memory chips that made possible small, affordable computing devices such as the personal computer. Similarly, the worth of MEMS may become apparent only when thousands or millions of mechanical structures are manufactured and integrated with electronic elements.

The first examples of mass production of microelectromechanical devices have begun to appear—and many others are being contemplated in research laboratories all over the world. An early prototype demonstrates how MEMS may affect the way millions of people spend their leisure time in front of the television set. Texas Instruments has built an electronic display in which the picture elements, or pixels, that make up the image are controlled by microelectromechanical structures. Each pixel consists of a 16-micron-wide aluminum mirror that can reflect pulses of colored light onto a screen. The pixels are turned off or on when an electric field causes the mirrors to tilt 10 degrees to one side or the other. In one direction, a light beam is reflected onto the screen to illuminate the pixel. In the other, it scatters away from the screen, and the pixel remains dark.

This micromirror display could project the images required for a large-screen television with a high degree of brightness and resolution of picture detail. The mirrors could compensate for the inadequacies encountered with other technologies. Display designers, for instance, have run into difficulty in making liquid-crystal screens large enough for a wall-size television display.

The future of MEMS can be glimpsed by examining projects that have been funded during the past three years under a program sponsored by the U.S. Department of Defense's Advanced Research Projects Agency. This research is directed toward building a number of prototype microelectromechanical devices and systems that could transform not only weapons systems but also consumer products.

A team of engineers at the University of California at Los

Angeles and the California Institute of Technology wants to show how MEMS may eventually influence aerodynamic design. The group has outlined its ideas for technology that might replace the relatively large moving surfaces of a wing—the flaps, slats and ailerons—that control both turning and ascent and descent. It plans to line the surface of a wing with thousands of 150-micron-long plates that, in their resting position, remain flat on the wing surface. When an electrical voltage is applied, the plates rise from the surface at up to a 90-degree angle. Thus activated, they can control the vortices of air that form across selected areas of the wing. Sensors can monitor the currents of air rushing over the wing and send a signal to adjust the position of the plates.

These movable plates, or actuators, function similarly to a microscopic version of the huge flaps on conventional aircraft. Fine-tuning the control of the wing surfaces would enable an airplane to turn more quickly, stabilize against turbulence, or burn less fuel because of greater flying efficiency. The additional aerodynamic control achieved with this "smart skin" could lead to radically new aircraft designs that move beyond the cylinder-with-wings appearance that has prevailed for 70 years. Aerospace engineers might dispense entirely with flaps, rudders and even the wing surface, called a vertical stabilizer. The aircraft would become a kind of "flying wing," similar to the U.S. Air Force's Stealth bomber. An aircraft without a vertical stabilizer would exhibit greater maneuverability—a boon for fighter aircraft and perhaps also one day for highspeed commercial airliners that must be capable of changing direction quickly to avoid collisions.

The engineering of small machines and sensors allows new uses for old ideas. For a decade, scientists have routinely worked with the scanning probe microscopes that can manipulate and form images with individual atoms. The most well known of these devices is the scanning tunneling microscope, or STM.

The STM, an invention for which Gerd Binnig and Heinrich Rohrer of IBM won the Nobel Prize in Physics in 1986, caught the attention of micromechanical specialists in the early 1980s. The fascination of the engineering community stems from calculations of how much information could be stored if STMs were used to read and write digital data. A trillion bits of information—equal to the text of 500 *Encyclopædia Britannicas*—might fit into a square centimeter on a chip by deploying an assembly of multiple STMs.

The STM is a needle-shaped probe, the tip of which consists of a single atom. A current that "tunnels" from the tip to a nearby conductive surface can move small groups of atoms, either to create holes or to pile up tiny mounds on the silicon chip. Holes and mounds correspond to the zeros and ones required to store digital data. A sensor, perhaps one constructed from a different type of scanning probe microscope, would "read" the data by detecting whether a nanometer-size plot of silicon represents a zero or one.

Only beams and motors a few microns in size, and with a commensurately small mass, will be able to move an STM quickly and precisely enough to make terabit (trillion-bit) data storage on a chip practicable. With MEMS, thousands of STMs could be suspended from movable beams built on the surface of a chip, each one reading or writing data in an area of a few square microns. The storage medium, moreover, could remain stationary, which would eliminate the need for today's spinning-media disk drives.

Noel C. MacDonald, an electrical engineering professor at Cornell University, has taken a step toward fulfilling the vision of the pocket research library. He has built an STM-equipped microbeam that can be moved in either the vertical and horizontal axes or even at an oblique angle. The beam hangs on a suspended frame attached to four motors, each of which measures only 200 microns (two hair widths) across. These engines push or pull on each side of the tip at speeds as high as a million times a second. MacDonald next plans to build an array of STMs.

The Lilliputian infrastructure afforded by MEMS might let chemists and biologists perform their experiments with instruments that fit in the palm of the hand. Westinghouse Science and Technology Center is in the process of reducing to the size of a calculator a 50-pound benchtop spectrometer, used for measuring the mass of atoms or molecules. A miniaturized mass spectrometer presages an era of inexpensive chemical detectors for do-it-yourself toxic monitoring.

In the same vein, Richard M. White, a professor at the University of California at Berkeley, contemplates a chemical factory on a chip. White has begun to fashion millimeter-diameter wells in a silicon chip, each of which holds a different chemical. An electrical voltage causes liquids or powders to move from the wells down a series of channels into a reaction chamber.

These materials are pushed there by micropumps made of piezoelectric materials that constrict and then immediately release sections of the channel. The snakelike undulations create a pumping motion. Once the chemicals are in the chamber, a heating plate causes them to react. An outlet channel from the chamber then pumps out what is produced in the reaction.

A pocket-calculator-size chemical factory could thus reconstitute freeze-dried drugs, perform DNA testing to detect water-borne pathogens or mix chemicals that can then be converted into electrical energy more efficiently than can conventional batteries. MEMS gives microelectronics an opening to the world beyond simply processing and storing information. Automobiles, scientific laboratories, televisions, airplanes and even the home medicine cabinet will never be the same.

Size means everything in a microprocessor, and the smaller you can make any component, the better it is. Now, chemists may have found a way to coax tiny wires to grow into microcircuits by using electrical fields.

From Chips to Cubes

W. Wayt Gibbs

I f you want to pack more circuitry into an electronic gadget—and in the world of electronic gadgets, more is almost always better—you have to use smaller wires. Engineers have two tools to do this, microsoldering and photolithography, both of which have proved phenomenally successful. But both are also pressing against known limits. To keep computer sophistication racing forward at its rocket sled pace, semiconductor outfits will need a fundamentally new way to build ever denser microcircuitry. Jean-Claude Bradley, a chemist at Drexel University in Philadelphia, thinks he is on to one. If his technique works as hoped, it might be used, decades from now, to make microprocessors that look more like cubes than chips.

The first step, however, is a much more modest one. Bradley and his colleagues created two copper wires to make an exceedingly simple circuit that lights up a tiny bulb. What is interesting is not so much what they did but what they did not do: they did not use any of the standard and experimental techniques for building circuitry. No robot-controlled soldering pens. No ultraviolet lamps or light-sensitive acid washes to etch micron-size wires. No marvelously detailed printing plates to stamp out a circuit pattern.

Bradley used only decidedly low-tech gear. "We start off with a project board just like you'd buy at Radio Shack," he says. The board is covered with a grid of holes, each hole capped by a copper ring. Bradley covered two adjacent rings with a single drop of water, then stuck platinum electrodes into the bottom of the holes so that they were close to, but did not contact, the rings. He plugged the electrodes into the rough equivalent of two nine-volt batteries. Almost immediately, a branch of copper began growing from one ring toward the other. Within 45 seconds, the wire completed the circuit.

"This is the first example of constructing circuitry simply by controlling an electrical field," Bradley asserts. "You don't need to touch the copper rings in any way." Indeed, in a paper published in *Nature*, Bradley reported that his lab has grown finer wires less than a micron thick—nearly as thin as the wires in computer chips—between copper particles floating freely in a solvent. But it will take much more work to create complex microcircuits using electrodeposition.

Bradley says electrochemists understand in rough terms why this process works. The voltage applied to the platinum electrodes creates an electrical field that surrounds the two copper rings. The field polarizes the copper: it forces positive charges to one side and negative charges to the other. The same thing happens to both rings, so if the two are side by side, the positive edge of one ring will face the negative edge of the other. Opposites attract, and in a strong field, the opposite edges can attract so strongly that the electrical force will rip copper atoms off one ring and dump them into the water-filled gulf between the two. Once enough copper atoms are in the water, they begin to coalesce into a solid wire, which grows until it contacts the other ring and creates a conduit that nullifies the voltage difference between the two rings.

That explains why the wires grow, but Bradley admits that many mysteries about the phenomenon will have to be solved before electrodeposition will yield useful circuits. The wires form branching, treelike structures, for example. Smooth wires conduct higher currents and higher frequency signals more readily. And computer logic is made from semiconductors such as silicon, as well as conductors such as aluminum. Bradley thinks he can probably make smooth semiconductor circuitry by using different materials and solvents and by strengthening the electrical field. But he has yet to prove this.

Perhaps more important, chemists still need to demonstrate what Bradley claims is "the technique's real potential: to construct truly three-dimensional circuits." Acid etches, soldering guns and printing plates work well only on flat surfaces; that is why microchips are so thin. But if metal particles are suspended within a porous cube, Bradley speculates, one could then use a mesh of electrodes or beams of polarized light to generate minute electrical fields and in this way to grow wires that run up and down as well as to and fro. Now that Drexel has applied for provisional patents, Bradley has begun looking for industrial partners to bankroll the next step in his research: to make circuits that are as tall as they are broad.

Many researchers recognize that if computers are to become much smaller and much faster, a whole new technology must replace what we are using now. Quantum computing, with the ability to employ a half-step gradation between the traditional 1 and 0 used in today's computers, could provide a key breakthrough.

ated into our environment. T
everywhere-in grocery check
nes, inside the workings of
mer direct the functioning of
tches, artificial hearts, an
tomotive assembly lines. Som
wly designed fighter jets ar
stable that without constant

Quantum-Mechanical Computers

Seth Lloyd

E very two years for the past 50, computers have become twice as fast while their components have become half as big. Circuits now contain wires and transistors that measure only one hundredth of a human hair in width. Because of this explosive progress, today's machines are millions of times more powerful than their crude ancestors. But explosions do eventually dissipate, and integrated-circuit technology is running up against its limits.

Advanced lithographic techniques can yield past $\frac{1}{100}$ the size of what is currently available. But at this scale—where bulk matter reveals itself as a crowd of individual atoms—integrated circuits barely function. A tenth the size again, the individuals assert their identity, and a single defect can wreak havoc. So if computers are to become much smaller in the future, new technology must replace or supplement what we now have.

Several decades ago pioneers such as Rolf Landauer and Charles H. Bennett, both at the IBM Thomas J. Watson Research Center, began investigating the physics of information-processing circuits, asking questions about where miniaturization might lead: How small can the components of circuits be made? How much energy must be used up in the course of computation? Because computers are physical devices, their basic operation is described by physics. One physical fact of life is that as the components of computer circuits become very small, their description must be given by quantum mechanics.

In the early 1980s Paul Benioff of Argonne National Laboratory built on Landauer and Bennett's earlier results to show that a computer could in principle function in a purely quantum-mechanical fashion. Soon after, David Deutsch of the Mathematical Institute at the University of Oxford and other scientists in the U.S. and Israel began to model quantum-mechanical computers to find out how they might differ from classical ones. In particular, they wondered whether quantum-mechanical effects might be exploited to speed computations or to perform calculations in novel ways.

By the middle of the decade, the field languished for several reasons. First, all these researchers had considered quantum computers in the abstract instead of studying actual physical systems—an approach that Landauer faulted on many counts. It also became evident that a quantum-mechanical computer might be prone to errors and have trouble correcting them. And apart from one suggestion, made by Richard Feynman of the California Institute of Technology, that quantum computers might be useful for simulating other quantum systems (such as new or unobserved forms of matter), it was unclear that they could solve mathematical problems any faster than their classical cousins.

In the past few years, the picture has changed. In 1993 I described a large class of familiar physical systems that might act as quantum computers in ways that avoid some of Landauer's objections. Peter W. Shor of AT&T Bell Laboratories has demonstrated that a quantum computer could be used to factor large numbers—a task that can foil the most powerful of conventional machines. And in 1995, workshops at the Institute for Scientific Interchange in Turin, Italy, spawned many designs for constructing quantum circuitry. More recently, H. Jeff Kimble's group at Caltech and David J. Wineland's team at the National Institute of Standards and Technology have built some of these prototype parts, whereas David Cory of the Massachusetts Institute of Technology and Isaac Chuang of Los Alamos National Laboratory have demonstrated simple versions of my 1993 design. This article explains how quantum computers might be assembled and describes some of the astounding things they could do that digital computers cannot.

Let's face it, quantum mechanics is weird. Niels Bohr, the Danish physicist who helped to invent the field, said, "Anyone who can contemplate quantum mechanics without getting dizzy hasn't properly understood it." For better or worse, quantum mechanics predicts a number of counterintuitive effects that have been verified experimentally again and again. To appreciate the weirdness of which quantum computers are capable, we need accept only a single strange fact called wave-particle duality.

Wave-particle duality means that things we think of as solid particles, such as basketballs and atoms, behave under some circumstances like waves and that things we normally describe as waves, such as sound and light, occasionally behave like particles. In essence, quantum-mechanical theory sets forth what kind of waves are associated with what kind of particles, and vice versa.

The first strange implication of wave-particle duality is that small systems such as atoms can exist only in discrete energy states. So when an atom moves from one energy state to another, it absorbs and emits energy in exact amounts, or "chunks," called photons, which might be considered the particles that make up light waves.

A second consequence is that quantum-mechanical waves,

like water waves, can be superposed, or added together. Taken individually, these waves offer a rough description of a given particle's position. When two or more such waves are combined, though, the particle's position becomes unclear. In some weird quantum sense, then, an electron can sometimes be both here and there at the same time. Such an electron's location will remain unknown until some interaction (such as a photon bouncing off the electron) reveals it to be either here or there but not both.

When two superposed quantum waves behave like one wave, they are said to be coherent; the process by which two coherent waves regain their individual identities is called decoherence. For an electron in a superposition of two different energy states (or, roughly, two different positions within an atom), decoherence can take a long time. Days can pass before a photon, say, will collide with an object as small as an electron, exposing its true position. In principle, basketballs could be both here and there at once as well (even in the absence of Michael Jordan). In practice, however, the time it takes for a photon to bounce off a ball is too brief for the eye or any instrument to detect. The ball is simply too big for its exact location to go undetected for any perceivable amount of time. Consequently, as a rule only small, subtle things exhibit quantum weirdness.

Quantum Information

Information comes in discrete chunks, as do atomic energy levels in quantum mechanics. The quantum of information is the bit. A bit of information is a simple distinction between two alternatives—no or yes, 0 or 1, false or true. In digital computers, the voltage between the plates in a capacitor represents a bit of information: a charged capacitor registers a 1 and an uncharged

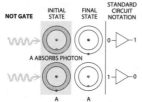

NOT GATE	INITIAL STATE	FINAL STATE	STANDARD CIRCUIT NOTATION

A ABSORBS PHOTON

A   A

NOT involves nothing more than bit flipping, as the notation above shows: if A is 0, make it a 1, and vice versa. With atoms, this can be done by applying a pulse whose energy equals the difference between A's ground state (its electron is in its lowest energy level, shown as the inner ring) and its excited state (shown as the outer ring). Unlike conventional NOT gates, quantum ones can also flip bits only halfway.

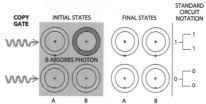

COPY, in the quantum world, relies on the interaction between two different atoms. Imagine one atom, A, storing either a 0 or 1, sitting next to another atom, B, in its ground state. The difference in energy between the states of B will be a certain value if A is 0, and another value if A is 1. Now apply a pulse of light whose photons have an energy equal to the latter amount. If the pulse is of the right intensity and duration and if A is 1, B will absorb a photon and flip (*top row*); if A is 0, B cannot absorb a photon from the pulse and stays unchanged (*bottom row*). So, as in the diagram below, if A is 1, B becomes 1; if A is 0, B remains 0.

COPY GATE	INITIAL STATES	FINAL STATES	STANDARD CIRCUIT NOTATION

B ABSORBS PHOTON

A   B   A   B

In the 19th century, Irish logician George Boole showed that all complex logical or arithmetic tasks could be accomplished using a combination of three simple operations: NOT, COPY, and AND. Any quantum system can perform these operations.

capacitor, a 0. A quantum computer functions by matching the familiar discrete character of digital information processing to the strange discrete character of quantum mechanics.

Indeed, a string of hydrogen atoms can hold bits as well as a string of capacitors. An atom in its electronic ground state could encode a 0 and in an excited state, a 1. For any such quantum system to work as a computer, though, it must be capable of more than storing bits. An operator must be able to load information onto the system, to process that information by way of simple logical manipulations and to unload it. That is, quantum systems must be capable of reading, writing and arithmetic.

Isidor Isaac Rabi, who was awarded the Nobel Prize for Physics in 1944, first showed how to write information on a quantum system. Applied to hydrogen atoms, his method works as follows. Imagine a hydrogen atom in its ground state, having an amount of energy equal to E_0. To write a 0 bit on this atom, do nothing. To write a 1, excite the atom to a higher energy level, E_1. We can do so by bathing it in laser light made up of photons having an amount of energy equal to the difference between E_1 and E_0. If the laser beam has the proper intensity and is

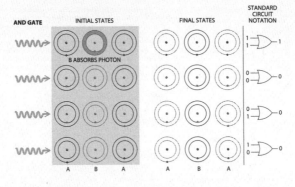

AND also depends on atomic interactions. Imagine three atoms, *A, B* and *A*, sitting next to one another. The difference in energy between the ground and excited states of *B* is a function of the states of the two *A*'s. Suppose *B* is in its ground state. Now apply a pulse whose energy equals the difference between the two states of *B* only when the atom's neighboring *A*'s are both 1. If, in fact, both *A*'s are 1, this pulse will flip *B* (*top row*); otherwise it will leave *B* unchanged (*all other rows*).

applied for the right length of time, the atom will gradually move from the ground state to the excited state, as its electron absorbs a photon. If the atom is already in the excited state, the same pulse will cause it to emit a photon and go to the ground state. In terms of information storage, the pulse tells the atom to flip its bit.

What is meant here by gradually? An oscillating electrical field such as laser light drives an electron in an atom from a lower energy state to a higher one in the same way that an adult pushes a child on a swing higher and higher. Each time the oscillating wave comes around, it gives the electron a little push. When the photons in the field have the same energy as the difference between E_0 and E_1, these pushes coincide with the electron's "swinging" motion and gradually convert the wave corresponding to the electron into a superposition of waves having different energies.

The amplitude of the wave associated with the electron's ground state will continuously diminish as the amplitude of the wave associated with the excited state builds. In the process, the bit registered by the atom "flips" from the ground state to the excited state. When the photons have the wrong frequency, their pushes are out of sync with the electron, and nothing happens.

If the right light is applied for half the time it takes to flip the atom from 0 to 1, the atom is in a state equal to a superposition of the wave corresponding to 0 and the wave corresponding to 1, each having the same amplitudes. Such a quantum bit, or qubit, is then flipped only halfway. In contrast, a classical bit will always read either 0 or 1. A half-charged capacitor in a conventional computer causes errors, but a half-flipped qubit opens the way to new kinds of computation.

Reading bits from a quantum system is similar to flipping them. Push the atom to an even higher, less stable energy state, call it E_2. Do so by subjecting the atom to light having an energy equal to the difference between E_1 and E_2: if the atom is in E_1, it will be excited to E_2 but decay rapidly back to E_1, emitting a photon. If the atom is already in the ground state, nothing happens. If it is in the "half-flipped" state, it has an equal chance of emitting a photon and revealing itself to be a 1 or of not emitting a photon, indicating that it is a 0. From writing and reading information in a quantum system, it is only a short step to computing.

Quantum Computation

Electronic circuits are made from linear elements (such as wires, resistors and capacitors) and nonlinear elements (such as diodes and transistors) that manipulate bits in different ways. Linear devices alter input signals individually. Nonlinear devices, on the other hand, make the input signals passing through them interact. If your stereo did not contain nonlinear transistors, for example, you could not change the bass in the music it plays. To do so requires some coordination of the information coming from your compact disc and the information coming from the adjustment knob on the stereo.

Circuits perform computations by way of repeating a few simple linear and nonlinear tasks over and over at great speed. One such task is flipping a bit, which is equivalent to the logical operation called NOT: true becomes false, and false becomes true. Another is COPY, which makes the value of a second bit the same as the first. Both those operations are linear, because in both the output reflects the value of a single input. Taking the AND of two bits—another useful task—is a nonlinear operation: if two input bits are both 1, make a third bit equal to 1 as well; otherwise make the third bit a 0. Here the output depends on some interaction between the inputs.

The devices that execute these operations are called logic gates. If a digital computer has linear logic gates, such as NOT and COPY gates, and nonlinear ones as well, such as AND gates, it can complete any logical or arithmetic task. The same requirements hold for quantum computers. Artur Ekert, working with Deutsch and Adriano Barenco at Oxford, and I have shown independently that almost any nonlinear interaction between quantum bits will do. Indeed, provided a quantum computer can flip bits, any nonlinear quantum interaction enables it to perform any computation. Hence, a variety of physical phenomena might be exploited to construct a quantum computer.

In fact, all-purpose quantum logic gates have been around almost as long as the transistor! In the late 1950s, researchers managed to perform simple two-bit quantum logic operations using particle spins. These spins—which are simply the orientation of a particle's rotation with respect to some magnetic field—are, like energy levels, quantized. So a spin in one direction can represent a 1 and in the other, a 0. The researchers took advantage of the interaction between the spin of the electron and the spin of the proton in a hydrogen atom; they set up a system in which they flipped the proton's spin only if the electron's spin represented a 1. Because these workers were not thinking about quantum logic, they called the effect double resonance. And yet they used double resonance to carry out linear NOT and COPY operations.

Since then, Barenco, David DiVincenzo of IBM, Tycho Sleator of New York University and Harald Weinfurter of the University of Innsbruck have demonstrated how, by flipping proton and electron spins only partway, double resonance can be used to create an AND gate as well. Such quantum logic gates, wired together, could make a quantum computer.

A number of groups have recently constructed quantum logic gates and proposed schemes for wiring them together. A particularly promising development has come from Caltech: by concentrating photons together with a single atom in a minute volume, Kimble's group has enhanced the usually tiny nonlinear interaction between photons. The result is a quantum logic gate: one photon bit can be flipped partway when another photon is in a state signifying 1. Quantum "wires" can be constructed by having single photons pass through optical fibers or through the air, in order to ferry bits of information from one gate to another.

An alternative design for a quantum logic circuit has been proposed by J. Ignacio Cirac of the University of Castilla-La Mancha in Spain and Peter Zoller of the University of Innsbruck. Their scheme isolates qubits in an ion trap, effectively insulating them from any unwanted external influences. Before a bit were processed, it would be transferred to a common register, or "bus." Specifically, the information it contained would be represented by a rocking motion involving all the ions in the trap. Wineland's group at NIST has taken the first step in realizing such a quantum computer, performing both linear and nonlinear operations on bits encoded by ions and by the rocking motion.

In an exciting theoretical development under experimental investigation at Caltech, Cirac, Zoller, Kimble and Hideo Mabuchi have shown how the photon and ion-trap schemes for quantum computing might be combined to create a "quantum Internet" in which photons are used to shuttle qubits coherently back and forth between distant ion traps.

Although their methods can in principle be scaled up to tens or hundreds of quantum bits, the Caltech and NIST groups have performed quantum logic operations on just two bits (leading some wags to comment that a two-bit micro-processor is just a two-bit microprocessor). In 1997, however, Neil A. Gershenfeld of M.I.T., together with Chuang of Los Alamos, showed that my 1993 method for performing quantum computing using the double resonance methods described above could be realized using nuclear spins at room temperature. The same result was obtained independently by M.I.T.'s Cory, working with Amr Fahmy and Timothy F. Havel of Harvard Medical School. With conventional magnets of the kind used to perform magnetic resonance imaging, Chuang and Cory both succeeded in performing quantum logic operations on three bits, with the prospect of constructing 10-bit quantum microprocessors in the near future.

Thus, as it stands, scientists can control quantum logic operations on a few bits, and in the near future, they might well do quantum computations using a few tens or hundreds of bits. How can this possibly represent an improvement over classical computers that routinely handle billions of bits? In fact, even with one bit, a quantum computer can do things no classical computer can. Consider the following. Take an atom in a superposition of 0 and 1. Now find out whether the bit is a 1 or a 0 by making it fluoresce. Half of the time, the atom emits a photon, and the bit is a 1. The other half of the time, no photon is emitted, and the bit is a 0. That is, the bit is a random bit—something a classical computer cannot create. The random-number programs in digital computers actually generate pseudorandom numbers, using a function whose output is so irregular that it seems to produce bits by chance.

Imagine what a quantum computer can do with two bits. Copying works by putting together two bits, one with a value to be copied and one with an original value of 0; an applied pulse flips the second bit to 1 only if the first bit is also 1. But if the value of the first bit is a superposition of 0 and 1, then the applied pulse creates a superposition involving both bits, such that both are 1 or both are 0. Notice that the final value of the first bit is no longer the same as it was originally—the superposition has changed.

In each component of this superposition, the second bit is the same as the first, but neither is the same as the original bit. Copying a superposition state results in a so-called entangled state, in which the original information no longer resides in a single quantum bit but is stored instead in the correlations between qubits. Albert Einstein noted that such states would violate all classical intuition about causality. In such a superposition, neither bit is in a definite state, yet if you measure one bit, thereby putting it in a definite state, the other bit also enters into a definite state. The change in the first bit does not *cause* the change in the second. But by virtue of destroying the coherence between the two, measuring the first bit also robs the second of its ambiguity. I have shown how quantum logic can be used to explore the properties of even stranger entangled states that involve correlations among three and more bits, and Chuang has used magnetic resonance to investigate such states experimentally.

Our intuition for quantum mechanics is spoiled early on in life. A one-year-old playing peekaboo knows that a face is there even when she cannot see it. Intuition is built up by manipulating objects over and over again; quantum mechanics seems counterintuitive because we grow up playing with classical toys. One of the best uses of quantum logic is to expand our intuition by allowing us to manipulate quantum objects and play with quantum toys such as photons and electrons.

The more bits one can manipulate, the more fascinating the phenomena one can create. I have shown that with more bits, a quantum computer could be used to simulate the behavior of any quantum system. When properly programmed, the computer's dynamics would become exactly the same as the dynamics of some postulated system, including that system's interaction with its environment. And the number of steps the computer would need to chart the evolution of this system over time would be directly proportional to the size of the system.

Even more remarkable, if a quantum computer had a parallel architecture, which could be realized through the exploitation of the double resonance between neighboring pairs of spins in the atoms of a crystal, it could mimic any quantum system in real time, regardless of its size. This kind of parallel quantum computation, if possible, would give a huge speedup over conventional methods. As Feynman noted, to simulate a quantum system on a classical computer generally requires a number of steps that rises exponentially both with the size of the system and with the amount of time over which the system's behavior is tracked. In fact, a 40-bit quantum computer could re-create in little more than, say, 100 steps, a quantum system that would take a classical computer, having a trillion bits, years to simulate.

What can a quantum computer do with many logical operations on many qubits? Start by putting all the input bits in an equal superposition of 0 and 1, each having the same magnitude. The computer then is in an equal superposition of all possible inputs. Run this input through a logic circuit that carries out a particular computation. The result is a superposition of all the possible outputs of that computation. In some weird quantum sense, the computer performs all possible computations at once. Deutsch has called this effect "quantum parallelism."

Quantum parallelism may seem odd, but consider how waves work in general. If quantum-mechanical waves were sound waves, those corresponding to 0 and 1—each oscillating at a single frequency—would be pure tones. A wave corresponding to a superposition of 0 and 1 would then be a chord. Just as a musical chord sounds qualitatively different from the individual tones it includes, a superposition of 0 and 1 differs from 0 and 1 taken alone: in both cases, the combined waves interfere with each other.

A quantum computer carrying out an ordinary computation, in which no bits are superposed, generates a sequence of waves analogous to the sound of "change ringing" from an English church tower, in which the bells are never struck simultaneously and the sequence of sounds follows mathematical rules. A computation in quantum-parallel mode is like a symphony: its "sound" is that of many waves interfering with one another.

Shor of Bell Labs has shown that the symphonic effect of quantum parallelism might be used to factor large numbers very quickly—something classical computers and even supercomputers cannot always accomplish. Shor demonstrated that a quantum-parallel computation can be orchestrated so that potential factors will stand out in the superposition the same way that a melody played on violas, cellos and violins an octave apart will stand out over the sound of the surrounding instruments in a symphony. Indeed, his algorithm would make factoring an easy task for a quantum computer, if one could be built. Because most public-key encryption systems—such as those protecting electronic bank accounts—rely on the fact that classical computers cannot find factors having more than, say, 100 digits, quantum-computer hackers would give many people reason to worry.

Whether or not quantum computers (and quantum hackers) will come about is a hotly debated question. Recall that the quantum nature of a superposition prevails only so long as the environment refrains from somehow revealing the state of the system. Because quantum computers might still consist of thousands or millions of atoms, only one of which need be disturbed to damage quantum coherence, it is not clear how long interacting quantum systems can last in a true quantum superposition. In addition, the various quantum systems that might be used to register and process information are susceptible to noise, which can flip bits at random.

Shor and Andrew Steane of Oxford have shown that quantum logic operations can be used to construct error-correcting routines that protect the quantum computation against decoherence and errors. Further analyses by Wojciech Zurek's group at Los Alamos and by John Preskill's group at Caltech have shown that quantum computers can perform arbitrarily complex computations as long as only one bit in 100,000 is decohered or flipped at each time step.

It remains to be seen whether the experimental precision required to perform arbitrarily long quantum computations can be attained. To surpass the factoring ability of current supercomputers, quantum computers using Shor's algorithm might need to follow thousands of bits over billions of steps. Even with the error correction, because of the technical problems described by Landauer, it will most likely prove rather difficult to build a computer to perform such a computation. To surpass classical simulations of quantum systems, however, would require only tens of bits followed for tens of steps, a more attainable goal. And to use quantum logic to create strange, multiparticle quantum states and to explore their properties is a goal that lies in our current grasp.

Unlike current microchips etched on a shard of silicon, the processing heart of a quantum computer could take the form of a liquid-filled capsule brimming with molecules. But to exploit the promise of a quantum-mechanical computer, several key hurdles must first be cleared.

Quantum Computing with Molecules

Neil Gershenfeld and Isaac L. Chuang

F actoring a number with 400 digits—a numerical feat needed to break some security codes—would take even the fastest supercomputer in existence billions of years. But a newly conceived type of computer, one that exploits quantum-mechanical interactions, might complete the task in a year or so, thereby defeating many of the most sophisticated encryption schemes in use. Sensitive data are safe for the time being, because no one has been able to build a practical quantum computer. But researchers have now demonstrated the feasibility of this approach. Such a computer would look nothing like the machine that sits on your desk; surprisingly, it might resemble the cup of coffee at its side.

· We and several other research groups believe quantum computers based on the molecules in a liquid might one day overcome many of the limits facing conventional computers. Roadblocks to improving conventional computers will ultimately arise from the fundamental physical bounds to miniaturization (for example, because transistors and electrical wiring cannot be made slimmer than the width of an atom). Or they may come about for practical reasons—most likely because the facilities for fabricating still more powerful microchips will become prohibitively expensive. Yet the magic of quantum mechanics might solve both these problems.

The advantage of quantum computers arises from the way they encode a bit, the fundamental unit of information. The state of a bit in a classical digital computer is specified by one number, 0 or 1. An n-bit binary word in a typical computer is accordingly described by a string of n zeros and ones. A quantum bit, called a qubit, might be represented by an atom in one of two different states, which can also be denoted as 0 or 1. Two qubits, like two classical bits, can attain four different well-defined states (0 and 0, 0 and 1, 1 and 0, or 1 and 1).

But unlike classical bits, qubits can exist simultaneously as 0 and 1, with the probability for each state given by a numerical coefficient. Describing a two-qubit quantum computer thus requires four coefficients. In general, n qubits demand 2^n numbers, which rapidly becomes a sizable set for larger values of n. For example, if n equals 50, about 10^{15} numbers are required to describe all the probabilities for all the possible states of the quantum machine—a number that exceeds the capacity of the largest conventional computer. A quantum computer promises to be immensely powerful because it can be in multiple states at once—a phenomenon called superposition—and because it can act on all its possible states simultaneously. Thus, a quantum computer could naturally perform myriad operations in parallel, using only a single processing unit.

Another property of qubits is even more bizarre—and useful. Imagine a physical process that emits two photons (packets of light), one to the left and the other to the right, with the two photons having opposite orientations (polarizations) for their oscillating electrical fields. Until detected, the polarization of each of the photons is indeterminate. As noted by Albert Einstein and others early in the century, at the instant a person measures the polarization for one photon, the state of the other polarization becomes immediately fixed—no matter how far away it is. Such instantaneous action at a distance is curious indeed. This phenomenon allows quantum systems to develop a spooky connection, a so-called entanglement, that effectively serves to wire together the qubits in a quantum computer. This same property allowed Anton Zeilinger and his colleagues at the University of Innsbruck in Austria to perform a remarkable demonstration of quantum teleportation in 1997.

In 1994 Peter W. Shor of AT&T deduced how to take advantage of entanglement and superposition to find the prime factors of an integer. He found that a quantum computer could, in principle, accomplish this task much faster than the best classical calculator ever could. His discovery had an enormous impact. Suddenly, the security of encryption systems that depend on the difficulty of factoring large numbers became suspect. And because so many financial transactions are currently guarded with such encryption schemes, Shor's result sent tremors through a cornerstone of the world's electronic economy.

Certainly no one had imagined that such a breakthrough would come from outside the disciplines of computer science or number theory. So Shor's algorithm prompted computer scientists to begin learning about quantum mechanics, and it sparked physicists to start dabbling in computer science.

The researchers contemplating Shor's discovery all understood that building a useful quantum computer was going to be fiendishly difficult. The problem is that almost any interaction a quantum system has with its environment—say, an atom colliding with another atom or a stray photon—constitutes a measurement. The superposition of quantum-mechanical states then collapses into a single very definite state—the one that is detected by an observer. This phenomenon, known as decoherence, makes further quantum calculation impossible. Thus, the inner workings of a quantum computer must somehow be separated from its surroundings to maintain coherence. But they must also be accessible so that calculations can be loaded, executed and read out.

Prior work, including elegant experiments by Christopher R. Monroe and David J. Wineland of the National Institute of Standards and Technology and by H. Jeff Kimble of the California Institute of Technology, attempted to solve this problem by carefully isolating the quantum-mechanical heart of their computers. For example, magnetic fields can trap a few charged particles, which can then be cooled into pure quantum states. But even such heroic experimental efforts have demonstrated only rudimentary quantum operations, because these novel devices involve only a few bits and because they lose coherence very quickly.

So how then can a quantum-mechanical computer ever be exploited if it needs to be so well isolated from its surroundings? Last year we realized that an ordinary liquid could perform all the steps in a quantum computation: loading in an initial condition, applying logical operations to entangled superpositions and reading out the final result. Along with another group at Harvard University and the Massachusetts Institute of Technology, we found that nuclear magnetic resonance (NMR) techniques (similar to the methods used for magnetic resonance imaging, or MRI) could manipulate quantum information in what appear to be classical fluids.

It turns out that filling a test tube with a liquid made up of appropriate molecules—that is, using a huge number of individual quantum computers instead of just one—neatly addresses the problem of decoherence. By representing each qubit with a vast collection of molecules, one can afford to let measurements interact with a few of them. In fact, chemists, who have used NMR for decades to study complicated molecules, have been doing quantum computing all along without realizing it.

Nuclear magnetic resonance operates on quantum particles in the atomic nuclei within the molecules of the fluid. Particles with "spin" act like tiny bar magnets and will line up with an externally applied magnetic field. Two alternative alignments (parallel or antiparallel to the external field) correspond to two quantum states with different energies, which naturally constitute a qubit. One might suppose that the parallel spin corresponds to the number 1 and the antiparallel spin to the number 0. The parallel spin has lower energy than the antiparallel spin, by an amount that depends on the strength of the externally applied magnetic field. Normally, opposing spins are present in equal numbers in a fluid. But the applied field favors the creation of parallel spins, so a tiny imbalance between the two states ensues. This minute excess, perhaps just one in a million nuclei, is measured during an NMR experiment.

In addition to this fixed magnetic backdrop, NMR procedures also utilize varying electromagnetic fields. By applying an oscillating field of just the right frequency (determined by the magnitude of the fixed field and the intrinsic properties of the particle involved), certain spins can be made to flip between states. This feature allows the nuclear spins to be redirected at will.

For instance, protons (hydrogen nuclei) placed within a fixed magnetic field of 10 tesla can be induced to change direction by a magnetic field that oscillates at about 400 megahertz—that is, at radio frequencies. While turned on, usually only for a few millionths of a second, such radio waves will rotate the nuclear spins about the direction of the oscillating field, which is typically arranged to lie at right angles to the fixed field. If the oscillating radio-frequency pulse lasts just long enough to rotate the spins by 180 degrees, the excess of magnetic nuclei previously aligned in parallel with the fixed field will now point in the opposite, antiparallel direction. A pulse of half that duration would leave the particles with an equal probability of being aligned parallel or antiparallel.

In quantum-mechanical terms, the spins would be in both states, 0 and 1, simultaneously. The usual classical rendition of this situation pictures the particle's spin axis pointing at 90 degrees to the fixed magnetic field. Then, like a child's top that is canted far from the vertical force of gravity, the spin axis of the particle itself rotates, or precesses, about the magnetic field, looping around with a characteristic frequency. In doing so, it emits a feeble radio signal, which the NMR apparatus can detect.

In fact, the particles in an NMR experiment feel more than just the applied fields, because each tiny atomic nucleus influences the magnetic field in its vicinity. In a liquid, the constant motion of the molecules relative to one another evens out most of these local magnetic ripples. But one magnetic nucleus can affect another in the same molecule when it disturbs the electrons orbiting around them both.

Rather than being a problem, this interaction within a molecule proves quite useful. It allows a logic "gate," the basic unit of a computation, to be readily constructed using two nuclear spins. For our two-spin experiments, we used chloroform ($CHCl_3$). We were interested in taking advantage of the interaction between the spins of the hydrogen and carbon nuclei. Because the nucleus of common carbon, carbon 12, has no spin, we used chloroform containing carbon with one extra neutron, which imparts an overall spin to it.

Suppose the spin of the hydrogen is directed either up or down, parallel or antiparallel to a vertically applied magnetic field, whereas the spin of the carbon is arranged so that it definitely points up, parallel to this fixed magnetic field. A properly designed radio-frequency pulse can rotate the carbon's spin downward into the horizontal plane. The carbon nucleus will then precess about the vertical, with a speed of rotation that depends on whether the hydrogen nucleus in that molecule also happens to be parallel to the applied field. After a certain short time, the carbon will point either in one direction or exactly the opposite, depending on whether the spin of the neighboring hydrogen was up or down. At that instant, we apply another radio-frequency pulse to rotate the carbon nucleus another 90 degrees. That maneuver then flips the carbon nucleus into the down position if the adjacent hydrogen was up or into the up position if the hydrogen was down.

This set of operations corresponds to what electrical engineers call an exclusive-OR logic gate, something that is perhaps better termed a controlled-NOT gate (because the state of one input controls whether the signal presented at the other input is inverted at the output). Whereas classical computers require similar two-input gates as well as simpler one-input NOT gates in their construction, a group of researchers showed in 1995 that quantum computations can indeed be performed by means of rotations applied to individual spins and controlled-NOT gates. In fact, this type of quantum logic gate is far more versatile than its classical equivalent, because the spins on which it is based can be in superpositions of up and down states. Quantum computation can therefore operate simultaneously on a combination of seemingly incompatible inputs.

In 1996 we set out with Mark G. Kubinec of the University of California at Berkeley to build a modest two-bit quantum-mechanical computer made from a thimbleful of chloroform. Preparing the input for even this two-bit device requires considerable effort. A series of radio-frequency pulses must transform the countless nuclei in the experimental liquid into a collection that has its excess spins arranged just right. Then these qubits must be sequentially modified. In contrast to the bits in a conventional electronic computer, which migrate in an orderly way through arrays of logic gates as the calculation proceeds, these qubits do not go anywhere. Instead the logic gates are brought to them using various NMR manipulations. In essence, the program to be executed is compiled into a series of radio-frequency pulses.

The first computation we accomplished that exercised the unique abilities of quantum-mechanical computing followed an ingenious search algorithm devised by Lov K. Grover of Bell Laboratories. A typical computer searching for a desired item that is lost somewhere in a database of n items would take, on average, about $n/2$ tries to find it. Amazingly, Grover's quantum search can pinpoint the desired item in roughly $\sqrt{n}$ tries. As an example of this savings, we demonstrated that our two-qubit quantum computer could find a marked item hidden in a list of four possibilities in a single step. The classical solution to this problem is akin to opening a two-bit padlock by guessing: one would be unlikely to find the right combination on the first attempt. In fact, the classical method of solution would require, on average, between two and three tries.

A basic limitation of the chloroform computer is clearly its small number of qubits. The number of qubits could be expanded, but n could be no larger than the number of atoms in the molecule employed. With existing NMR equipment, the biggest quantum computers one can construct would have only about 10 qubits (because at room temperature the strength of the desired signal decreases rapidly as the number of magnetic nuclei in the molecule increases). Special NMR instrumentation designed around a suitable molecule could conceivably extend that number by a factor of three or four. But to create still larger computers, other techniques, such as optical pumping, would be needed to "cool" the spins. That is, the light from a suitable laser could help align the nuclei as effectively as removing the thermal motion of the molecules— but without actually freezing the liquid and ruining its ability to maintain long coherence times.

So larger quantum computers might be built. But how fast would they be? The effective cycle time of a quantum computer is determined by the slowest rate at which the spins flip around. This rate is, in turn, dictated by the interactions between spins and typically ranges from hundreds of cycles a second to a few cycles a second. Although running only a handful of clock cycles each second might seem awfully sluggish compared with the megahertz speed of conventional computers, a quantum computer with enough qubits would achieve such massive quantum parallelism that it would still factor a 400-digit number in about a year.

Given such promise, we have thought a great deal about how such a quantum computer could be physically constructed. Finding molecules with enough atoms is not a problem. The frustration is that as the size of a molecule increases, the interactions between the most distant spins eventually become too weak to use for logic gates. Yet all is not lost. Seth Lloyd of M.I.T. has shown that powerful quantum computers could, in principle, be built even if each atom interacts with only a few of its nearest neighbors, much like today's parallel computers. This kind of quantum computer might be made of long hydrocarbon molecules, also using NMR techniques. The spins in the many atomic nuclei, which are linked into long chains, would then serve as the qubits.

Another barrier to practical NMR computation is coherence. Rotating nuclei in a fluid will, like synchronized swimmers robbed of proper cues, begin to lose coherence after an interval of a few seconds to a few minutes. The longest coherence times for fluids, compared with the characteristic cycle times, suggest that about 1,000 operations could be performed while still preserving quantum coherence. Fortunately, it is possible to extend this limit by adding extra qubits to correct for quantum errors.

Although classical computers use extra bits to detect and correct errors, many experts were surprised when Shor and others showed that the same can be done quantum-mechanically. They had naively expected that quantum error correction would

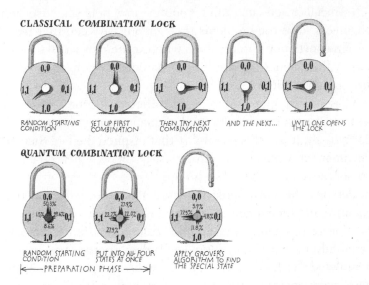

CLASSICAL COMBINATION LOCK

RANDOM STARTING CONDITION   SET UP FIRST COMBINATION   THEN TRY NEXT COMBINATION   AND *THE NEXT...*   UNTIL ONE OPENS THE LOCK

QUANTUM COMBINATION LOCK

RANDOM STARTING CONDITION   PUT INTO ALL FOUR STATES AT ONCE   APPLY GROVER'S ALGORITHM TO FIND THE SPECIAL STATE

|←——— PREPARATION PHASE ———→|

Cracking a combination lock requires fewer tries with some quantum wizardry. For example, a two-bit classical lock might demand as many as four attempts to open it (top). On average, an n-bit lock requires about $n/2$ tries. Because a quantum lock can be put into multiple states at once, it takes only about $\sqrt{n}$ steps to open it if Grover's algorithm is used.

require measuring the state of the system and hence wrecking its quantum coherence. It turns out, however, that quantum errors can be corrected within the computer without the operator ever having to read the erroneous state.

Still, reaching sizes that make quantum computers large enough to compete with the fastest classical computers will be especially difficult. But we believe the challenge is well worth taking on. Quantum computers, even modest ones, will provide superb natural laboratories in which to study the principles of quantum mechanics. With these devices, researchers will be able to investigate other quantum systems that are of fundamental interest simply by running the appropriate program.

Ironically, such quantum computers may help scientists and engineers solve the problems they encounter when they try to design conventional microchips with exceedingly small transistors, which behave quantum-mechanically when reduced in size to their limits.

Classical computers have great difficulty solving such problems of quantum mechanics. But quantum computers might do so easily. It was this possibility that inspired the late Richard Feynman of Caltech to ponder early on whether quantum computers could actually be built.

Perhaps the most satisfying aspect is the realization that constructing such quantum computers will not require the fabrication of tiny circuits of atomic scale or any other sophisticated advance in nanotechnology. Indeed, nature has already completed the hardest part of the process by assembling the basic components. All along, ordinary molecules have known how to do a remarkable kind of computation. People were just not asking them the right questions.

Maintaining the fragile quantum state needed to make quantum computers function is a major obstacle in quantum computing. Light waves may be an ideal medium for one class of quantum computers that exploits the wave nature of light.

crochips have seamlessly in
ated into our environment. T
everywhere-in grocery check
nes, inside the workings of
meras, hwith of ca
ey direct the functioning of
tches, artificial hearts, an
tomotive assembly lines. Som
wly designed fighter jets ar
stable that without constant

Computing with Light

Graham P. Collins

L arge quantum computers could in principle handle some of the toughest computing problems, such as factoring numbers to break encrypted messages—answering those questions in seconds instead of the centuries that today's computers would require. But quantum computers are extraordinarily difficult to build; they rely on exquisitely controlled interactions among fragile quantum states. Do they have to? Recently Ian A. Walmsley and his co-workers at the University of Rochester demonstrated that ordinary, classical light waves can perform as efficiently as one class of quantum computer.

The Rochester experiment searched a sorted 50-element database. An ordinary computer doing a binary search of such a database would need to query the database six times (enough to search 64 elements: $2^6 = 64$). In 1997 Lov K. Grover of Bell Laboratories proved that a quantum computer only has to query once, no matter how large the database.

Double Quantum Dot Computing

Purdue University physicist Albert Chang and colleagues have successfully linked two so-called quantum dots such that the tiny structures could conceivably serve as qubits—switches for quantum computers that can be on, off or in a combination of states. So that the pairs of would-be qubits could exchange information, the scientists connected quantum dots, essentially flat pools of electrons measuring a mere 180-nanometers wide, using even finer nanowires made by way of standard electron beam lithography. But they had to do so in such a way that they could still determine which state each individual dot was in, depending on its electron spin. "Each dot can have a one [on] or a zero [off] because the spin can be up or down," Chang explains. But "without being able to isolate each spin, you cannot do quantum computation."

The trick, the team reports, lay in controlling how many electrons ended up in each quantum dot. As in atoms, the electrons fill the dot by taking up positions in successive orbitals around the center. Chang and colleagues took advantage of the fact that when there was only one electron in the dot's outermost orbital, they could analyze the flow of electricity through the dots and thereby detect spin. "Now we have proven that you can link quantum dots together," Chang says, "but the next thing will be to make them do things, to control the spins in a double-quantum dot."
—*Kristin Leutwyler*

Walmsley's group used a light pulse in an interferometer, a device that gives light waves a choice of two paths to follow. Along one path, a diffraction grating splits the pulse apart into its broad range of frequencies, like white light through a prism. The 50 elements of the database correspond to 50 bands of that spectrum. The database itself is represented by an acousto-optic modulator through which the light passes. The modulator imprints a phase shift (that is, it moves the positions of the peaks and troughs of the light wave) on just one of the 50 bands. In essence, each band of the light "looks at" a different database entry (a different part of the modulator), and only one "finds" the target. When the pulse is recombined with light from the other arm of the interferometer, the phase-shifted band alone shines brightly into a spectrometer, which reads off the result. Only the wave nature of light, not its quantum features, is used.

The experiment is similar to established methods of optical signal processing that, for example, pass beams through holograms. What's new is that it directly exemplifies a general result that Walmsley and his colleagues demonstrated theoretically late last year. "For every machine that uses [only] quantum interference," Walmsley explains, "there is an equivalent, equally efficient one that uses classical optical interference." Reading out a result on a quantum computer necessarily involves detection of particles, and the extra device components and computational steps for that process eliminate the quantum computer's advantage. According to Emanuel H. Knill of Los Alamos National Laboratory, that insight provides a new perspective "on the relationship between computing with waves and quantum computing."

The most powerful quantum algorithms, such as fast factoring, however, require an additional quantum feature: so-called entanglement of the states of many particles. Classical waves cannot emulate those algorithms efficiently, but light turns out to be well suited to such true quantum computation in another way. In theory, a full-power quantum computer can be built by sending individual photons through simple linear optical elements, such as beam splitters and phase shifters. Such an approach was proposed in 1997, but those early designs needed exponentially more optical elements as the number of qubits increased—utterly impractical for any but the smallest devices.

Recently, Knill, his colleague Raymond Laflamme and Gerard J. Milburn of the University of Queensland in Australia exhibited a design whose circuit complexity would increase in linear proportion, not exponentially. Unlike the Rochester experiment, this scheme relies on quantum effects of individual photons navigating paths through the device but avoids the need for nonlinear interactions between photons, something only readily achieved at very high intensities or with extraordinary equipment such as resonant cavities or light-slowing Bose-Einstein condensates.

Necessity is the mother of invention. Many research centers, hamstrung by budget cuts, simply couldn't afford million dollar supercomputers. Their solution was to build their own—out of discarded PCs from the office workplace. Clusters of outdated PCs working together have produced near supercomputer performance thanks to parallel computing, a technique where complex tasks are divided into smaller problems for processing.

The Do-It-Yourself Supercomputer

William W. Hargrove, Forrest M. Hoffman and Thomas Sterling

I n the well-known stone soup fable, a wandering soldier stops at a poor village and says he will make soup by boiling a cauldron of water containing only a shiny stone. The townspeople are skeptical at first but soon bring small offerings: a head of cabbage, a bunch of carrots, a bit of beef. In the end, the cauldron is filled with enough hearty soup to feed everyone. The moral: cooperation can produce significant achievements, even from meager, seemingly insignificant contributions.

Researchers are now using a similar cooperative strategy to build supercomputers, the powerful machines that can perform billions of calculations in a second. Most conventional supercomputers employ parallel processing: they contain arrays of ultrafast microprocessors that work in tandem to solve complex problems such as forecasting the weather or simulating a nuclear explosion. Made by IBM, Cray and other computer vendors, the machines typically cost tens of millions of dollars—far too much for a research team with a modest budget. So over the past few years, scientists at national laboratories and universities have learned how to construct their own supercomputers by linking inexpensive PCs and writing software that allows these ordinary computers to tackle extraordinary problems.

In 1996 two of us (Hargrove and Hoffman) encountered such a problem in our work at Oak Ridge National Laboratory (ORNL) in Tennessee. We were trying to draw a national map of ecoregions, which are defined by environmental conditions: all areas with the same climate, landforms and soil characteristics fall into the same ecoregion. To create a high-resolution map of the continental U.S., we divided the country into 7.8 million square cells, each with an area of one square kilometer. For each cell we had to consider as many as 25 variables, ranging from average monthly precipitation to the nitrogen content of the soil. A single PC or workstation could not accomplish the task. We needed a parallel-processing supercomputer—and one that we could afford!

Our solution was to construct a computing cluster using obsolete PCs that ORNL would have otherwise discarded. Dubbed the Stone SouperComputer because it was built essentially at no cost, our cluster of PCs was powerful enough to produce ecoregion maps of unprecedented detail. Other research groups have devised even more capable clusters that rival the performance of the world's best supercomputers at a mere fraction of their cost. This advantageous price-to-performance ratio has already attracted the attention of some corporations, which plan to use the clusters for such complex tasks as deciphering the human genome. In fact, the cluster concept promises to revolutionize the computing field by offering tremendous processing power to any research group, school or business that wants it.

The notion of linking computers together is not new. In the 1950s and 1960s the U.S. Air Force established a network of vacuum-tube computers called SAGE to guard against a Soviet nuclear attack. In the mid-1980s Digital Equipment Corporation coined the term "cluster" when it integrated its mid-range VAX minicomputers into larger systems. Networks of workstations—generally less powerful than minicomputers but faster than PCs—soon became common at research institutions. By the early 1990s scientists began to consider building clusters of PCs, partly because their mass-produced microprocessors had become so inexpensive. What made the idea even more appealing was the falling cost of Ethernet, the dominant technology for connecting computers in local-area networks.

Advances in software also paved the way for PC clusters. In the 1980s Unix emerged as the dominant operating system for scientific and technical computing. Unfortunately, the operating systems for PCs lacked the power and flexibility of Unix. But in 1991 Finnish college student Linus Torvalds created Linux, a Unix-like operating system that ran on a PC. Torvalds made Linux available free of charge on the Internet, and soon hundreds of programmers began contributing improvements. Now wildly popular as an operating system for stand-alone computers, Linux is also ideal for clustered PCs.

The first PC cluster was born in 1994 at the NASA Goddard Space Flight Center. NASA had been searching for a cheaper way to solve the knotty computational problems typically encountered in earth and space science. The space agency needed a machine that could achieve one gigaflops—that is, perform a billion floating-point operations per second. (A floating-point operation is equivalent to a simple calculation such as addition or multiplication.) At the time, however, commercial supercomputers with that level of performance cost about $1 million, which was too expensive to be dedicated to a single group of researchers.

One of us (Sterling) decided to pursue the then radical concept of building a computing cluster from PCs. Sterling and his Goddard colleague Donald J. Becker connected 16 PCs, each containing an Intel 486 microprocessor, using Linux and a standard Ethernet network. For scientific applications, the PC cluster delivered sustained performance of 70 megaflops—that is, 70 million floating-point operations per second. Though modest by today's standards, this speed was not much lower than that of some smaller commercial supercomputers available at the time. And the cluster was built for only $40,000, or about one tenth the price of a comparable commercial machine in 1994.

NASA researchers named their cluster Beowulf, after the lean, mean hero of medieval legend who defeated the giant monster Grendel by ripping off one of the creature's arms. Since then, the name has been widely adopted to refer to any low-cost cluster constructed from commercially available PCs. In 1996 two successors to the original Beowulf cluster appeared: Hyglac (built by researchers at the California Institute of Technology and the Jet Propulsion Laboratory) and Loki (constructed at Los Alamos National Laboratory). Each cluster integrated 16 Intel Pentium Pro microprocessors and showed sustained performance of over one gigaflops at a cost of less than $50,000, thus satisfying NASA's original goal.

The Beowulf approach seemed to be the perfect computational solution to our problem of mapping the ecoregions of the U.S. A single workstation could handle the data for only a few states at most, and we couldn't assign different regions of the country to separate workstations—the environmental data for every section of the country had to be compared and processed simultaneously. In other words, we needed a parallel-processing system. So in 1996 we wrote a proposal to buy 64 new PCs containing Pentium II microprocessors and construct a Beowulf-class supercomputer. Alas, this idea sounded implausible to the reviewers at ORNL, who turned down our proposal.

Undeterred, we devised an alternative plan. We knew that obsolete PCs at the U.S. Department of Energy complex at Oak Ridge were frequently replaced with newer models. The old PCs were advertised on an internal Web site and auctioned off as surplus equipment. A quick check revealed hundreds of outdated computers waiting to be discarded this way. Perhaps we could build our Beowulf cluster from machines that we could collect and recycle free of charge. We commandeered a room at ORNL that had previously housed an ancient mainframe computer. Then we began collecting surplus PCs to create the Stone SouperComputer.

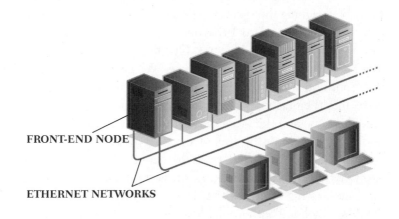

FRONT-END NODE

ETHERNET NETWORKS

Parallel computing systems start with a front-end node that directs communication. Problems are solved by dividing the computational workload into many tasks that are assigned to the nodes or individual computers in the group connected via an Ethernet network.

The strategy behind parallel computing is "divide and conquer." A parallel-processing system divides a complex problem into smaller component tasks. The tasks are then assigned to the system's nodes—for example, the PCs in a Beowulf cluster—which tackle the components simultaneously. The efficiency of parallel processing depends largely on the nature of the problem. An important consideration is how often the nodes must communicate to coordinate their work and to share intermediate results. Some problems must be divided into myriad minuscule tasks; because these fine-grained problems require frequent internode communication, they are not well suited for parallel processing. Coarse-grained problems, in contrast, can be divided into relatively large chunks. These problems do not require much communication among the nodes and therefore can be solved very quickly by parallel-processing systems.

Anyone building a Beowulf cluster must make several decisions in designing the system. To connect the PCs, researchers can use either standard Ethernet networks or faster, specialized networks, such as Myrinet. Our lack of a budget dictated that we use Ethernet, which is free. We chose one PC to be the front-end node of the cluster and installed two Ethernet cards into the machine. One card was for communicating with outside users, and the other was for talking with the rest of the nodes, which would be linked in their own private network. The PCs coordinate their tasks by sending messages to one another. The two most popular message-passing libraries are message-passing interface (MPI) and parallel virtual machine (PVM), which are both available at no cost on the Internet. We use both systems in the Stone SouperComputer.

Many Beowulf clusters are homogeneous, with all the PCs containing identical components and microprocessors. This uniformity simplifies the management and use of the cluster but is not an absolute requirement. Our Stone SouperComputer would have a mix of processor types and speeds because we intended to use whatever surplus equipment we could find. We began with PCs containing Intel 486 processors but later added only Pentium-based machines with at least 32 megabytes of RAM and 200 megabytes of hard-disk storage.

It was rare that machines met our minimum criteria on arrival; usually we had to combine the best components from several PCs. We set up the digital equivalent of an automobile thief's chop shop for converting surplus computers into nodes for our cluster. Whenever we opened a machine, we felt the same anticipation that a child feels when opening a birthday present: Would the computer have a big disk, lots of memory or (best of all) an upgraded motherboard donated to us by accident? Often all we found was a tired old veteran with a fan choked with dust.

Our room at Oak Ridge turned into a morgue filled with the picked-over carcasses of dead PCs. Once we opened a machine, we recorded its contents on a "toe tag" to facilitate the extraction of its parts later on. We developed favorite and least favorite brands, models and cases and became adept at thwarting passwords left by previous owners. On average, we had to collect and process about five PCs to make one good node.

As each new node joined the cluster, we loaded the Linux operating system onto the machine. We soon figured out how to eliminate the need to install a keyboard or monitor for each node. We created mobile "crash carts" that could be wheeled over and plugged into an ailing node to determine what was wrong with it. Eventually someone who wanted space in our room bought us shelves to consolidate our collection of hardware. The Stone SouperComputer ran its first code in early 1997, and by May 2001 it contained 133 nodes, including 75 PCs with Intel 486 microprocessors, 53 faster Pentium-based machines and five still faster Alpha workstations, made by Compaq.

Upgrades to the Stone SouperComputer are straightforward: we replace the slowest nodes first. Each node runs a simple speed test every hour as part of the cluster's routine housekeeping tasks. The ranking of the nodes by speed helps us to fine-tune our cluster. Unlike commercial machines, the performance of the Stone SouperComputer continually improves, because we have an endless supply of free upgrades.

* * *

Parallel programming requires skill and creativity and may be more challenging than assembling the hardware of a Beowulf system. The most common model for programming Beowulf clusters is a master-slave arrangement. In this model, one node acts as the master, directing the computations performed by one or more tiers of slave nodes. We run the same software on all the machines in the Stone SouperComputer, with separate sections of code devoted to the master and slave nodes. Each microprocessor in the cluster executes only the appropriate section. Programming errors can have dramatic effects, resulting in a digital train wreck as the crash of one node derails the others. Sorting through the wreckage to find the error can be difficult.

Another challenge is balancing the processing workload among the cluster's PCs. Because the Stone SouperComputer contains a variety of microprocessors with very different speeds, we cannot divide the workload evenly among the nodes: if we did so, the faster machines would sit idle for long periods as they waited for the slower machines to finish processing. Instead we developed a programming algorithm that allows the master node to send more data to the faster slave nodes as they complete their tasks. In this load-balancing arrangement, the faster PCs do most of the work, but the slower machines still contribute to the system's performance.

Our first step in solving the ecoregion mapping problem was to organize the enormous amount of data—the 25 environmental characteristics of the 7.8 million cells of the continental U.S. We created a 25-dimensional data space in which each dimension represented one of the variables (average temperature, precipitation, soil characteristics and so on). Then we identified each cell with the appropriate point in the data space. Two points close to each other in this data space have, by definition, similar characteristics and thus are classified in the same ecoregion. Geographic proximity is not a factor in this kind of classification; for example, if two mountaintops have very similar environments, their points in the data space are very close to each other, even if the mountaintops are actually thousands of miles apart.

Once we organized the data, we had to specify the number of ecoregions that would be shown on the national map. The cluster of PCs gives each ecoregion an initial "seed position" in the data space. For each of the 7.8 million data points, the system determines the closest seed position and assigns the point to the corresponding ecoregion. Then the cluster finds the centroid for each ecoregion—the average position of all the points assigned to the region. This centroid replaces the seed position as the defining point for the ecoregion. The cluster then repeats the procedure, reassigning the data points to ecoregions depending on their distances from the centroids. At the end of each iteration, new centroid positions are calculated for each ecoregion. The process continues until fewer than a specified number of data points change their ecoregion assignments. Then the classification is complete.

The mapping task is well suited for parallel processing because different nodes in the cluster can work independently on subsets of the 7.8 million data points. After each iteration the slave nodes send the results of their calculations to the master node, which averages the numbers from all the subsets to determine the new centroid positions for each ecoregion. The master node then sends this information back to the slave nodes for the next round of calculations. Parallel processing is also useful for selecting the best seed positions for the ecoregions at the very beginning of the procedure. We devised an algorithm that allows the nodes in the Stone SouperComputer to determine collectively the most widely dispersed data points, which are then chosen as the seed positions. If the cluster starts with well-dispersed seed positions, fewer iterations are needed to map the ecoregions.

The result of all our work was a series of maps of the continental U.S. showing each ecoregion in a different color. We produced maps showing the country divided into as few as four ecoregions and as many as 5,000. The maps with fewer ecoregions divided the country into recognizable zones—for example, the Rocky Mountain states and the desert Southwest. In contrast, the maps with thousands of ecoregions are far more complex than any previous classification of the country's environments. Because many plants and animals live in only one or two ecoregions, our maps may be useful to ecologists who study endangered species.

In our first maps the colors of the ecoregions were randomly assigned, but we later produced maps in which the colors of the ecoregions reflect the similarity of their respective environments. We statistically combined nine of the environmental variables into three composite characteristics, which we represented on the map with varying levels of red, green and blue. When the map is drawn this way, it shows gradations of color instead of sharp borders: the lush Southeast is mostly green, the cold Northeast is mainly blue, and the arid West is primarily red.

Moreover, the Stone SouperComputer was able to show how the ecoregions in the U.S. would shift if there were nationwide changes in environmental conditions as a result of global warming. Using two projected climate scenarios developed by other research groups, we compared the current ecoregion map with the maps predicted for the year 2099. According to these projections, by the end of this century the environment in Pittsburgh will be more like that of present-day Atlanta, and conditions in Minneapolis will resemble those in present-day St. Louis.

* * *

The traditional measure of supercomputer performance is benchmark speed: how fast the system runs a standard program. As scientists, however, we prefer to focus on how well the system can handle practical applications. To evaluate the Stone SouperComputer, we fed the same ecoregion mapping problem to ORNL's Intel Paragon supercomputer shortly before it was retired. At one time, this machine was the laboratory's fastest, with a peak performance of 150 gigaflops. On a per-processor basis, the run time on the Paragon was essentially the same as that on the Stone SouperComputer. We have never officially clocked our cluster (we are loath to steal computing cycles from real work), but the system has a theoretical peak performance of about 1.2 gigaflops. Ingenuity in parallel algorithm design is more important than raw speed or capacity: in this young science, David and Goliath (or Beowulf and Grendel!) still compete on a level playing field.

The Beowulf trend has accelerated since we built the Stone SouperComputer. New clusters with exotic names—Grendel, Naegling, Megalon, Brahma, Avalon, Medusa and the Hive, to mention just a few—have steadily raised the performance curve by delivering higher speeds at lower costs. As of November 2000, clusters of PCs, workstations or servers were on the list of the world's 500 fastest computers. The Los Lobos cluster at the University of New Mexico has 512 Intel Pentium III processors and is the 80th-fastest system in the world, with a performance of 237 gigaflops. The Cplant cluster at Sandia National Laboratories has 580 Compaq Alpha processors and is ranked 84th. The National Science Foundation and the U.S. Department of Energy are planning to build even more advanced clusters that could operate in the teraflops range (one trillion floating-point operations per second), rivaling the speed of the fastest supercomputers on the planet.

Beowulf systems are also muscling their way into the corporate world. Major computer vendors are now selling clusters to businesses with large computational needs. IBM, for instance, is building a cluster of 1,250 servers for NuTec Sciences, a biotechnology firm that plans to use the system to identify disease-causing genes. An equally important trend is the development of networks of PCs that contribute their processing power to a collective task. An example is SETI@home, a project launched by researchers at the University of California at Berkeley who are analyzing deep-space radio signals for signs of intelligent life. SETI@home sends chunks of data over the Internet to more than three million PCs, which process the radio-signal data in their idle time. Some experts in the computer industry predict that researchers will eventually be able to tap into a "computational grid" that will work like a power grid: users will be able to obtain processing power just as easily as they now get electricity.

Above all, the Beowulf concept is an empowering force. It wrests high-level computing away from the privileged few and makes low-cost parallel-processing systems available to those with modest resources. Research groups, high schools, colleges or small businesses can build or buy their own Beowulf clusters, realizing the promise of a supercomputer in every basement.

Fluid dynamics is a mathematically complex science, and many of today's supercomputers spend a great deal of time tracing and mapping these dynamic, variable flows. And as the speed of supercomputers increases, we are gaining insight into the puzzling nature of turbulence.

ated into our environment. T
everywhere-in grocery check
nes, inside the workings of
meras, undercarriages of ca
ey direct the functioning of
tches, artificial hearts, an
tomotive assembly lines. Som
wly designed fighter jets ar
stable that without constant

Tackling Turbulence with Supercomputers

Parviz Moin and John Kim

W e all pass through life surrounded—and even sustained—by the flow of fluids. Blood moves through the vessels in our bodies, and air (a fluid, properly speaking) flows into our lungs. Our vehicles move through our planet's blanket of air or across its lakes and seas, powered by still other fluids, such as fuel and oxidizer, that mix in the combustion chambers of engines. Indeed, many of the environmental or energy-related issues we face today cannot possibly be confronted without detailed knowledge of the mechanics of fluids.

Practically all the fluid flows that interest scientists and engineers are turbulent ones; turbulence is the rule, not the exception, in fluid dynamics. A solid grasp of turbulence, for example, can allow engineers to reduce the aerodynamic drag on an automobile or a commercial airliner, increase the maneuverability of a jet fighter or improve the fuel efficiency of an engine. An understanding of turbulence is also necessary to comprehend the flow of blood in the heart, especially in the left ventricle, where the movement is particularly swift.

But what exactly is turbulence? A few everyday examples may be illuminating. Open a kitchen tap only a bit, and the water that flows from the faucet will be smooth and glassy. This flow is known as laminar. Open the tap a little further, and the flow becomes more roiled and sinuous—turbulent, in other words. The same phenomenon can be seen in the smoke streaming upward into still air from a burning cigarette. Immediately above the cigarette, the flow is laminar. A little higher up, it becomes rippled and diffusive.

Turbulence is composed of eddies: patches of zigzagging, often swirling fluid, moving randomly around and about the overall direction of motion. Technically, the chaotic state of fluid motion arises when the speed of the fluid exceeds a specific threshold, below which viscous forces damp out the chaotic behavior.

Turbulence, however, is not simply an unfortunate phenomenon to be eliminated at every opportunity. Far from it: many engineers work hard trying to increase it. In the cylinders of an internal-combustion engine, for example, turbulence enhances the mixing of fuel and oxidizer and produces cleaner, more efficient combustion. And only turbulence can explain why a golf ball's dimples enable a skilled golfer to drive the ball 250 meters, rather than 100 at most.

Turbulence may have gotten its bad reputation because dealing with it mathematically is one of the most notoriously thorny problems of classical physics. For a phenomenon that is literally ubiquitous, remarkably little of a quantitative nature is known about it. Richard Feynman, the great Nobel Prize-winning physicist, called turbulence "the most important unsolved problem of classical physics." Its difficulty was wittily expressed in 1932 by the British physicist Horace Lamb, who, in an address to the British Association for the Advancement of Science, reportedly said, "I am an old man now, and when I die and go to heaven there are two matters on which I hope for enlightenment. One is quantum electrodynamics, and the other is the turbulent motion of fluids. And about the former I am rather optimistic."

Of course, Lamb could not have foreseen the development of the modern supercomputer. These technological marvels are at last making it possible for engineers and scientists to gain fleeting but valuable insights into turbulence. Already this work has led to technology, now in development, that may someday be employed on airplane wings to reduce drag by several percent—enough to save untold billions of dollars in fuel costs. At the same time, these insights are guiding the design of jet engines to improve both efficiency and performance.

As recondite as it is, the study of turbulence is a major component of the larger field of fluid dynamics, which deals with the motion of all liquids and gases. Similarly, the application of powerful computers to simulate and study fluid flows that happen to be turbulent is a large part of the burgeoning field of computational fluid dynamics (CFD). In recent years, fluid dynamicists have used supercomputers to simulate flows in such diverse cases as the America's Cup racing yachts and blood movement through an artificial heart.

What do we mean when we speak of simulating a fluid flow on a computer? In simplest terms, the computer solves a series of well-known equations that are used to compute, for any point in space near an object, the velocity and pressure of the fluid flowing around that object. These equations were discovered independently more than a century and a half ago by the French engineer Claude Navier and the Irish mathematician George Stokes. The equations, which derive directly from Newton's laws of motion, are known as the Navier-Stokes equations. It was the application of supercomputers to these equations that gave rise to the field of computational fluid dynamics; this marriage has been one of the greatest achievements in fluid dynamics since the equations themselves were formulated.

Although the marriage has been successful, the courtship was a rather long one. Not until the late 1960s did supercomputers begin achieving processing rates fast enough to solve the Navier-Stokes equations for some fairly straightforward cases, such as two-dimensional, slowly moving flows about an obstacle. Before then, wind tunnels were essentially the only way of testing the aerodynamics of new aircraft designs. Even today the limits of the most powerful supercomputers still make it necessary to resort to wind tunnels to verify the design for a new airplane.

Although both computational fluid dynamics and wind tunnels are now used for aircraft development, continued advances in computer technology and algorithms are giving CFD a bigger share of the process. This is particularly true in the early design stages, when engineers are establishing key dimensions and other basic parameters of the aircraft. Trial and error dominate this process, and wind-tunnel testing is very expensive, requiring designers to build and test each successive model. Because of the increased role of computational fluid dynamics, a typical design cycle now involves between two and four wind-tunnel tests of wing models instead of the 10 to 15 that were once the norm.

Another advantage of supercomputer simulations is, ironically, their ability to simulate more realistic flight conditions. Wind-tunnel tests can be contaminated by the influence of the tunnel's walls and the structure that holds the model in place. Some of the flight vehicles of the future will fly at many times the speed of sound and under conditions too extreme for wind-tunnel testing. For hypersonic aircraft (those that will fly at up to 20 times the speed of sound) and spacecraft that fly both within and beyond the atmosphere, computational fluid dynamics is the only viable tool for design. For these vehicles, which pass through the thin, uppermost levels of the atmosphere, nonequilibrium air chemistry and molecular physics must be taken into account.

Engine designers also rely extensively on computational techniques, particularly in the development of jet engines. A program called Integrated High Performance Turbine Engine Technology is seeking a 100 percent improvement in the thrust-to-weight ratio of jet engines and a 40 percent improvement in fuel efficiency by 2003. The project is supported by the U.S. Department of Defense, the National Aeronautics and Space Administration and various makers of jet engines.

The flow of air and fuel through a jet engine's various sections and passages is complex. A fan draws air into an internal chamber called a compressor. There multiple rotating and stationary stages increase the pressure about 20-fold. This high-pressure air is fed into a combustor, where it mixes with fuel and is ignited. Finally, the hot, very expanded exhaust drives a turbine. This turbine powers the fan and the compressor and, more important, generates thrust by directing the exhaust out of the rear of the engine at high velocity. Currently engineers use computational fluid dynamics to design turbine blades, inlet passages and the geometry of combustors. Simulations also help engineers shape the afterburner mixers that, in military aircraft, provide additional thrust for greater maneuverability. And they play a role in designing nacelles, the bulbous, cylindrical engine casings that typically hang below the wings.

To understand how the Navier-Stokes equations work, consider the flow of air over an airplane in flight. In reality, it will probably be many decades before computers are powerful enough to simulate in a detailed manner the fluid flows over an entire airplane. In theory, however, the Navier-Stokes equations can reveal the velocity and pressure of the air rushing by any point near the aircraft's surface. Engineers could then use these data to compute, for various flight conditions, all aerodynamic parameters of interest—namely, the lift, drag and moments (twisting forces) exerted on the airplane.

Drag is particularly important because it determines an aircraft's fuel efficiency. Fuel is one of the largest operating expenses for most airlines. Not surprisingly, aircraft companies have spent huge sums to reduce drag by even tiny increments. In general, though, lift is relatively easy to calculate, moments are harder, and drag is hardest of all.

Drag is difficult to compute mainly because it is the parameter most dependent on turbulence. Of course, in this context we are not referring to the bumpiness that provokes the pilot to remind passengers to fasten their seat belts. Even when a plane is flying smoothly, the flow of air within a few centimeters of its surface, in a volume known as the boundary layer, is turbulent. Because of turbulence, the high-speed air several millimeters above the surface of the wings is brought very close to the surface, where it undergoes a more abrupt—and momentum-robbing—deceleration. The equal and opposite reaction to this flow deceleration is drag on the aircraft. A great deal of the work of aerodynamicists involves understanding the mechanics of the generation and destruction of turbulence well enough to control it.

To solve the Navier-Stokes equations, engineers start by entering into the equations certain variables known as initial and boundary conditions. For an airplane in flight, the initial conditions include wind velocity and atmospheric disturbances, such as air currents. The boundary conditions include the precise shape of the aircraft, translated into mathematical coordinates.

Before the equations can be applied to an aircraft, computer specialists must represent the aircraft's surface and the space around it in a form usable by the computer. So they represent the airplane and its surroundings as a series of regularly spaced points, known as a computational grid. They then supply the coordinates and related parameters of the grid to the software that applies the Navier-Stokes equations to the data. The computer calculates a value for the parameters of interest—air velocity and pressure—for each of the grid points.

In effect, the computational grid breaks up (the technical term is "discretizes") the computational problem in space; the calculations are carried out at regular intervals to simulate the passage of time, so the simulation is temporally discrete as well. The closer together—and therefore more numerous—the points are in the computational grid, and the more often they are computed (the shorter the time interval), the more accurate and realistic the simulation is. In fact, for objects with complex shapes, even defining the surface and generating a computational grid can be a challenge.

Unfortunately, entering the initial and boundary conditions does not guarantee a solution, at least not with the computers available today or in the foreseeable future. The difficulty arises from the fact that the Navier-Stokes equations are nonlinear; in other words, the many variables in the equations vary with respect to one another by powers of two or greater. Interaction of these nonlinear variables generates a broad range of scales, which can make solving the equations exceedingly difficult. Specifically, in turbulence, the range of the size of whirling eddies can vary 1,000-fold or even more. There are other complicating factors as well, such as global dependence: the nature of the equations is that the fluid pressure at one point depends on the flow at many other points. Because the different parts of the problem are so interrelated, solutions must be obtained at many points simultaneously.

Although the preceding description conveys the basics of a fluid dynamics simulation, it leaves out turbulence, without which a realistic discussion of the capabilities—and limitations—of computational fluid dynamics would be futile. The complexities engendered by turbulence severely limit our ability to simulate fluid flow realistically.

Perhaps the simplest way to define turbulence is by reference to the Reynolds number, a parameter that compactly characterizes a flow. Named after the British engineer Osborne

Reynolds, this number indicates the ratio, or relative importance, of the flow's inertial forces to its viscous ones. (A flow's inertial force is calculated by multiplying together the fluid's density and the square of its velocity and dividing this product by a characteristic length of the flow, such as the width of the airfoil, if the flow is over a wing.)

Large inertial forces, relative to the viscous ones, tend to favor turbulence, whereas high viscosity staves it off. Put another way, turbulence occurs when the Reynolds number exceeds a certain value. The number is proportional to both the size of the object and the flow velocity. For example, the Reynolds number for air flowing over the fuselage of a cruising commercial aircraft is in the neighborhood of 100 million. For the air flowing past a good fastball, the Reynolds number is about 200,000. For blood flowing in a midsize artery, it is about 1,000.

As we have seen, a distinguishing characteristic of a turbulent flow is that it is composed of eddies, also known as vortices, in a broad range of sizes. These vortices are continually forming and breaking down. Large eddies break down into smaller ones, which break down into yet smaller eddies, and so on. When eddies become small enough, they simply dissipate viscously into heat. The British meteorologist Lewis F. Richardson described this process in verse:

Big whorls have little whorls,
Which feed on their velocity,
And little whorls have lesser whorls,
And so on to viscosity.

To solve the Navier-Stokes equations for, say, the flow over an airplane requires a finely spaced computational grid to resolve the smallest eddies. On the other hand, the grid must be large enough to encompass the entire airplane and some of the space around it. The disparity of length scales in a turbulent flow—the ratio of largest to smallest eddy size—can be calculated by raising the flow's Reynolds number to the ¾ power. This ratio can be used to estimate the number of grid points that are needed for a reasonably accurate simulation: because there are three dimensions, the number is proportional to the cube of this ratio of length scales. Thus, the required number of grid points for a numerical simulation is proportional to the Reynolds number raised to the ⁹/₄ power. In other words, doubling the Reynolds number results in almost a five-fold increase in the number of points required in the grid to simulate the flow.

Consider a transport airplane with a 50-meter-long fuse-lage and wings with a chord length (the distance from the leading to the trailing edge) of about five meters. If the craft is cruising at 250 meters per second at an altitude of 10,000 meters, about 10 quadrillion (10^{16}) grid points are required to simulate the turbulence near the surface with reasonable detail.

What kind of computational demands does this number of points impose? A rough estimate, based on current algorithms and software, indicates that even with a supercomputer capable of performing a trillion (10^{12}) floating-point operations per second, it would take several thousand years to compute the flow for one second of flight time! Such a "teraflops" computer does not yet exist, although researchers are now attempting to build one at Sandia National Laboratories. It will be about 10 times faster than the most powerful systems available today.

Fortunately, researchers need not simulate the flow over an entire aircraft to produce useful information. Indeed, doing so would probably generate much more data than we would know what to do with. Typically, fluid dynamicists care only about the effects of turbulence on quantities of engineering significance, such as the mean flow of a fluid or, in the case of an aircraft, the drag and lift forces and the transfer of heat. In the case of an engine, designers may be interested in the effects of turbulence on the rates at which fuel and oxidizer mix.

The Navier-Stokes equations are therefore often averaged over the scales of the turbulence fluctuations. What this means is that, in practice, researchers rarely calculate the motion of each and every small eddy. Instead they compute the large eddies and then use ad hoc modeling practices to estimate the effects of the small eddies on the larger ones. This practice gives rise to a simulated averaged flow field that is smoother than the actual flow—and thus drastically reduces the number of grid points necessary to simulate the field.

The ad hoc models that this averaging process demands range in complexity from simple enhanced coefficients of viscosity to entire additional systems of equations. All these models require some assumptions and contain adjustable coefficients that are derived from experiments. Therefore, at present, simulations of averaged turbulent flows are only as good as the models they contain.

As computers become more powerful, however, fluid dynamicists are finding that they can directly simulate greater proportions of turbulent eddies, enabling them to reduce the range of scales that are modeled. These approaches are a compromise between a direct numerical simulation of turbulence, in which all scales of motion are resolved, and the turbulence-averaged computations.

For years, meteorologists have used a form of this strategy called large-eddy simulation for weather prediction. In meteorology, the large-scale turbulent motions are of particular interest, so in meteorological applications the relatively large eddies are generally simulated in their entirety. Smaller-scale eddies are important only inasmuch as they may affect the larger-scale turbulence, so they are merely modeled. Recently engineers have begun using these techniques for simulating complex fluid flows, such as the gases inside a cylinder of an internal-combustion engine.

Another current trend in computational fluid dynamics, also made possible by increasing computational speed, is the direct, complete simulation of relatively simple flows, such as the flow in a pipe. Simple as they are, simulations of some of these flows, which have low Reynolds numbers, offer invaluable insights into the nature of turbulence. They have revealed the basic structure of turbulent eddies near a wall and subtleties of their influence on drag. They have also generated useful data that have enabled engineers to validate or fine-tune the ad hoc models they use in practical simulations of complex flows.

Lately the number of engineers and scientists seeking access to these data has swelled to the point that immense data sets have been archived and made available by the NASA Ames Research Center. Although most researchers do not have the computing resources to perform direct simulations of turbulence, they do have sufficient resources, such as powerful workstations, to probe the archived data.

As supercomputers become faster and faster, fluid dynamicists are increasingly able to move beyond predicting the effects of turbulence to actually controlling them. Such control can have enormous financial benefits; a 10 percent reduction in the drag of civilian aircraft, for example, could yield a 40 percent increase in the profit margin of an airline. In a recent project, researchers at the NASA Langley Research Center demonstrated that placing longitudinal V-shaped grooves, called riblets, on the surface of an aircraft's wing or fuselage leads to a 5 to 6 percent reduction in viscous drag. Drag is reduced despite the increase in the surface area exposed to the flow. For typical transport airplane speeds, the riblets must be very finely spaced, about 40 microns apart, like phonograph grooves; larger riblets tend to increase drag.

During this work, the researchers came across a Soviet study on toothlike structures, called denticles, on the skin of sharks. These denticles strikingly resembled riblets, a fact that has been interpreted as nature's endorsement of the riblet concept. Ultimately, however, it was the direct numerical simulation of turbulent flow along riblets that showed how they work. The riblets appear to inhibit the motion of eddies by preventing them from coming very close to the surface (within about 50 microns). By keeping the eddies this tiny distance away, the riblets prevent the eddies from transporting high-speed fluid close to the surface, where it decelerates and saps the aircraft's momentum.

Another recent and exciting application of direct numerical simulation is in the development of turbulence-control strategies that are active (as opposed to such passive strategies as riblets). With these techniques the surface of, say, a wing would be moved slightly in response to fluctuations in the turbulence of the fluid flowing over it. The wing's surface would be built with composites having millions of embedded sensors and actuators that respond to fluctuations in the fluid's pressure and speed in such a way as to control the small eddies that cause the turbulence drag.

Such technology appeared to be farfetched six years ago, but the advent of microelectromechanical systems (MEMS), under the leadership of the U.S. Air Force, has brought such a scheme to the brink of implementation. MEMS technology can fabricate integrated circuits with the necessary microsensors, control logic and actuators. Active control of turbulence near the wing's surface also has an encouraging analogue in the form of a marine creature: dolphins achieve remarkable propulsive efficiency as they swim, and fluid dynamicists have long speculated that these creatures do it by moving their skins. It seems that smart aircraft skins, too, are endorsed by nature.

Getting back to those golf balls we mentioned earlier: they, too, present an intriguing example of how a surface texture can advantageously control airflow. The most important drag exerted on a golf ball derives from air-pressure forces. This phenomenon arises when the air pressure in front of the ball is significantly higher than the pressure behind the ball. Because of the turbulence generated by the dimples, a golf ball is able to fly about two and a half times farther than an identical but undimpled ball.

The growing popularity of computational fluid dynamics to study turbulence reflects both its promise, which is at last starting to be realized, and the continued rapid increase in computational power. As supercomputer processing rates approach and surpass a trillion floating-point operations per second over the next few years, fluid dynamicists will begin taking on more complex turbulent flows, of higher Reynolds numbers. Over the next decade, perhaps, researchers will simulate the flow of air through key passages in a jet engine and obtain a realistic simulation of an operating piston-cylinder assembly in an internal-combustion engine, including the intake and combustion of fuel and the exhaust of gases through valves. Through simulations such as these, researchers will finally begin learning some of the deep secrets expressed by the equations uncovered by Navier and Stokes a century and a half ago.

A next-generation hardware platform and software refinement in current supercomputing designs could lead supercomputers to their next evolutionary step: hypercomputers. Featuring state-of-the-art components bathed in chilled, rarefied environments, hypercomputers could achieve speeds 100 times greater than any existing supercomputer.

ated 'into our environment. 1
everywhere-in grocery check
nes, inside the workings of
meras, the heads of ca
ey direct the functioning of
tches, artificial hearts, ar
tomotive assembly lines. Son
wly designed fighter jets ar
stable that without constant

How to Build a Hypercomputer

Thomas Sterling

Today's fastest supercomputers run too slowly to do tomorrow's science. Despite the ongoing revolution in communications and information processing, many computational challenges critical to the future health, welfare, security and prosperity of humankind cannot be met by even the quickest computers. Crucial advances in pivotal fields such as climatology, medicine, bioscience, controlled fusion, national defense, nanotechnology, advanced engineering and commerce depend on the development of machines that will operate at speeds at least 1,000 times faster than today's biggest supercomputers.

Solutions to these incredibly complex problems hinge on the ability to simulate and model their behavior with a high degree of fidelity and reliability, often over long periods. This level of performance goes far beyond that of present-day supercomputers, which at best can execute several trillion floating-point operations per second (teraflops). It could take 100 years, for example, for the largest existing system to perform a complete protein-folding computation—a long-sought capability. To accomplish this kind of analysis task, researchers need hypercomputing systems that achieve at least petaflops speeds—that is, more than a quadrillion floating-point operations (arithmetical calculations) per second.

Not only do current high-end computers run too slowly, they cost too much. The three-teraflops (peak performance) ASCI (Accelerated Strategic Computer Initiative) Blue systems that are dedicated to the stewardship of the U.S. nuclear stockpile cost approximately $120 million each. That's equivalent to a price/performance factor of $40 per peak megaflops (million flops), which is more than 10 times greater than the price/performance of a premium personal computer. High-end computers impose indirect costs as well. Annual payments for the electrical power to operate such systems can easily exceed $1 million. Housing their oversize footprints can also add significant expense. Paying crack programmers to write the complex code for these machines is yet another cost.

Despite their impressive processing speeds, high-end systems do not make good use of the computing resources they have, resulting in surprisingly low efficiency levels. Twenty-five percent efficiency is not uncommon, and efficiencies have dropped as low as 1 percent when addressing certain applications.

The hybrid technology multithreaded (HTMT) system is a new class of computer that offers 100 times the capability of present high-end machines for roughly the same cost, power usage and floor space. Further development could bring the technology beyond a quadrillion flops to trans-petaflops territory—1,000 times the performance of today's best systems or more. To achieve these goals, a multi-institutional, interdisciplinary team has created a computer architecture able to harness various advanced processing, memory and communications technologies, leveraging their strengths and complementing their limitations. The basic elements of HTMT have been developed with financial support from NASA, the National Security Agency, the National Science Foundation and the Defense Advanced Research Projects Agency; actual construction awaits further governmental funding.

Ironically, it is the very success of computing technology that reveals its limitations. Back in the late 1970s, personal computers could barely play Pong. A system capable of executing a major science problem of the day at a performance level of a few tens of megaflops could cost $40 million or more. In contrast, PCs now priced at less than $2,000 can outperform those machines.

Historically, the supercomputer industry has pushed the frontiers of processing performance with a combination of advanced technology and architectures customized to address specific problems. The unfortunate side effect has been high price tags. Exorbitant costs and lengthy development times have kept the market for such systems relatively flat while other segments of the computer industry have grown explosively. With these costs forcing up the price to the customer, the overall supercomputer market and corporate investment in the technology have remained limited, producing a classic commercial death spiral.

Even when alternative approaches have been tried—including custom vector computer architectures (which efficiently perform a single operation on a list of numbers using pipelined memory access and arithmetic functional units) as well as massively parallel systems integrating large arrays of cooperating microprocessors—the costs of such systems have remained high while operational efficiencies for many applications have suffered. In the past four or five years, a number of groups have built highly parallel general-purpose computers with peak performance levels of more than teraflops. Yet low efficiency levels mean that little of this processing capability can be brought to bear on real-world applications. As a result, commodity clusters—networked arrays of standard computing subsystems—

are perceived as the only economically viable pathway: They require little additional development in spite of the programming difficulties and communications delays inherent in using clustered systems.

Research on new classes of petaflops-capable systems has been under way since the mid-1990s. Engineers have been attacking the speed problem on all fronts, pursuing various technology paths to such machines. With sufficient R&D support, all can be accomplished within this decade. Although each method has its strengths and weaknesses, one of the most widely applicable is the HTMT design.

HTMT exploits a diverse array of advanced technologies within a single flexible and optimized system. The project attempts to achieve efficient trans-petaflops performance by incorporating superfast processors, high-capacity communications links, high-density memory storage and other soon-to-mature technologies in a dynamic, adaptive architecture.

No matter what course they take, designers of trans-petaflops systems all face three challenges. First, they must find a way to aggregate sufficient processing, memory and communications resources to achieve the targeted peak-computing capabilities despite practical constraints of size, cost and power. The second goal is to attain reasonable operational efficiencies in the face of standard degradation factors. These include latencies (time delays) across the system, contention for shared resources such as common memory and communications channels, overhead-related resource reductions caused by the need to manage and coordinate concurrent tasks and parallel resources, and wastage of computing resources (starvation) caused by insufficient task parallelism or inadequate load balancing. The third objective concerns finding ways to improve the usability of the system—a somewhat arbitrary measure comprising the issues of generality (general utility), programmability and availability.

*   *   *

Crucial Tasks for Hypercomputers

Many intricate scientific problems with enormous social and political implications await solutions that can be processed only on computers that can execute more than a quadrillion floating-point operations per second—trans-petaflops performance.

Climate Modeling
Perhaps the most critical issue facing the earth's inhabitants is the need for accurate predictive scenarios for both short- and long-term weather changes. First, trans-petaflops computers could integrate the huge quantities of satellite data into detailed maps. The mapped data could then be used to simulate and model the chaotic and interrelated behaviors of the elements of our global climate system, allowing accurate predictions.

Controlled Fusion
Both an answer to the world's energy problems and a way to power spacecraft across the solar system, thermonuclear fusion's vast complexity has kept it continually just over the horizon. Trans-petaflops computers would simulate the thermal, electromagnetic and nuclear interactions of large numbers of particles in a dynamic magnetic medium to help in designing practical fusion reactors.

Medicine/Bioscience
Considerably faster computing capability could give medicine the edge in combating continuously evolving diseases. This job requires molecular-level analysis to achieve nearly instantaneous drug design, including exploring complex protein folding.

Agriculture
To feed the earth's ever growing population, rapid computation will help develop new genetically engineered crops and solve the complex problems involved in managing the world's ecology.

National Defense
With the real-world testing of nuclear weapons banned, transpetaflops machines could model the behavior of these systems to help maintain the readiness of the strategic weapons stockpile. Real-time decryption of increasingly complicated secret codes is one key to maintaining national defense.

Commerce and Finance
Large-scale mining of the enormous data spaces containing business information and economic statistics will allow a more accurate simulation of commercial systems.

Nanotechnology
With digital electronics shrinking to the atomic scale, where quantum mechanics is important, chip designers can no longer model electronics using averaged physical parameters.

Advanced Engineering
Ultrafast processing will be needed to simulate the behavior of new materials and composites at the microscale. Future aircraft design and that of other complex engineered systems will benefit from the same type of detailed modeling capabilities.

Astronomy
To model the galaxy and its 100 billion stars properly, new, superfast computers will be required to analyze the complex interplay of the interstellar medium and heavier molecules.

During the past decade, digital logic has been dominated by CMOS (complementary metal oxide semiconductor) processors. CMOS technology has provided lower power and greater performance while system densities have increased at an exponential rate. Yet the fastest digital logic technology on Earth is not CMOS. An altogether different technology using another kind of physics claims that title: superconducting logic.

Discovered at the beginning of the 20th century, superconductivity is the ability to conduct electricity with no resistance, a phenomenon that some materials exhibit when cooled to cryogenic temperatures. In principle, a loop of superconducting wire can sustain an electric current forever. More important, superconducting devices exhibit quantum-mechanical behavior in macroscale electronic components and circuits. In the early 1960s researchers developed a nonlinear switching device based on superconductivity called the Josephson junction, which was found to have exceptional speeds.

The HTMT hypercomputer design will employ high-speed superconducting logic processors based on Josephson junction technology. In rapid single-flux quantum (RSFQ) technology, superconducting loops store information as tiny magnetic flux quanta (by discrete current levels). The loops, called superconducting quantum interference devices, or SQUIDs, are simple mechanisms originally developed as sensing devices that comprise two Josephson junctions connected by an inductor, which is like a solenoid. With both Josephson junctions operating, a current injected into the loop will continue indefinitely. SQUIDs exhibit the interesting characteristic of having distinct states of operation: They may contain no current, sustain the basic current, or have a current that is some integral multiple times the basic current but nothing in between. This remarkable property results from quantum-mechanical effects. To represent the 0's and 1's of digital code, RSFQ logic gates use discrete currents (or fluxes) rather than distinct voltage levels. When cooled to a temperature of four kelvins, these units can operate at more than 770 gigahertz, the fastest (single-gate) processing speeds ever achieved and approximately 100 times quicker than conventional CMOS logic.

RSFQ technology will allow the hybrid computing system to run nominally at from 100 to 200 gigaflops (billion flops) per processor as opposed to a few gigaflops, as in standard CMOS processors. In addition, the minuscule and packetized nature of magnetic flux quanta in RSFQ devices cuts crosstalk and power consumption by a couple of orders of magnitude. This rapidly maturing technology reduces parallelism requirements, cost, power demand and system size.

With superfast processors in place, HTMT seeks to make efficient use of their powerful capabilities. Those processors should spend their time doing little else but computations. Conventional approaches such as commodity clusters require large-scale tasks to be run on similarly large-scale computational nodes. Often a computational node on a conventional system must wait while a remote request to another node is being serviced. Unless operators exactly balance the workload, some nodes will continue to compute while others, having finished their jobs, will stall. Even when engineers employ load-balancing software techniques, the overhead required for accomplishing this function can reduce efficiency.

Unlike any other computer architecture, HTMT revolutionizes the relation between the processing system and the memory system. In ordinary multiprocessor systems, the computational processors manage and manipulate the "dumb" memory system; in contrast, HTMT's "smart" memory system administers the processors. HTMT and other tightly coupled parallel computers consider the workload on the processing elements and make on-the-fly decisions as to which part of a task should be performed by what hardware. In doing so, the processors work out of their local registers and some high-

Five Routes to Ultrafast Processing

One approach to attaining trans-petaflops computing performance (more than a quadrillion floating-point operations per second) is to use a hybrid architecture combining several soon-to-be-available advanced technologies (*see accompanying article*). Here are five other technical pathways to achieving that goal.

Name	Method	Example	Best Applications
Special Purpose Architecture or Systolic Array	Specially designed hardware and software that mirror the abstract problem to be solved. Runs in parallel with a fast data pipeline to speed computation.	GRAPE Project (University of Tokyo)	Huge multibody calculations stellar cluster simulation bioinformatics
Cellular Automata	Finite-state machine in which many relatively simple computing cells in a large 2-D or 3-D matrix operate in step during each clock cycle. Each cell's actions depend on its internal state and those of its nearest neighbors.	Never fully executed	Computational fluid dynamics diffusion, simulations
Processor-in-Memory (PIM) Architecture	With a good deal of processing and memory on each chip, the system logic sees all the bits coming out of the dynamic random-access memory (DRAM) at the same time. There's lots of memory access and little delay in data transmission speed processing during each cycle.	IRAM (University of California at Berkeley) Blue Gene (IBM)	Image processing data encryption, rapid database searches, protein-folding modeling
Beowulf or Cluster Architectures	A high-bandwidth mesh system interconnects many low-cost, commodity processors (each a partial-system-on-a-chip device) in a high-density array.	GigAssembler software (International Human Genome Sequencing Consortium)	Wide range of problems; deciphering the human genome
Distributed Computing or Megacomputing Architectures	Harness the unused computing cycles on the estimated 500 million personal computers linked to the Internet. Inefficient communications are a drawback.	SETI@home (Serendip Project)	Huge parallel problems such as Monte Carlo simulations and monitoring the function of the Internet

speed buffer memories, thus avoiding having to reach too far out into the system. The result is a drop in latency problems. The processors do not spend time managing memory resources, which are just wasted processing cycles that add to overhead; these logistical decisions are made by the small low-cost processors in the memory.

The HTMT design attacks the problem of latency in two ways. First, the system employs a dynamic, adaptive resource management scheme based on a multithreaded architecture that enables HTMT to switch from one stream of instructions to another within a single cycle. Whereas most computers operate with one stream of instructions, HTMT will feature multiple instruction streams. By using overlapping communications, the processors can work on many outstanding requests simultaneously. Say a superconducting processor needs to load information from a cache or a high-speed buffer, a procedure that will take many 10-picosecond cycles. As this request is served by the memory system, the processor can switch to another data stream to find operations that can be performed immediately.

The second way HTMT will handle the latency issue is by employing processor-in-memory technology (PIM), wherein small secondary satellite or taxi logic processors are placed in its memory devices. A few years ago fabrication advances allowed CMOS logic and dynamic random-access memory (DRAM) cells to be put onto the same silicon die, permitting them to be closely integrated. These cheap devices deal with overhead—that is, they manipulate information in the memory, again allowing the superconducting processors to focus on computing. PIM processing technology can also handle memory-intensive functions such as data gathers—collecting needed information from various locations and placing it in one dense object—as well as carrying out the reverse operation of data scatters—distributing the information to the correct locations.

Although the technologies and architecture incorporated by the HTMT system may be innovative, the means of managing these resources and the computing discipline employed by HTMT are truly revolutionary. The system will use new percolation techniques in which the PIM processors decide when a new piece of work should be performed. They will determine when to migrate all information that needs to be executed up to rapid-access buffer memories near the high-speed superconducting processors. For example, when a specific subroutine is required, it and the special information it needs to execute its function will be moved up to the processors. This proactive method of prestaging necessary information is a way to avoid creating long latency delays in connecting to the main memory. The technique also frees the high-speed processors from having to perform logistical overhead operations, because they are not needed to bring information to the processing sites.

The third major challenge to trans-petaflops computing concerns the usability of the system: Researchers must increase its generality (to ensure that it can handle a wide variety of problems), make it easier to program, and boost its availability, or uptime. HTMT addresses these issues in several ways.

By using a global name space in a shared-memory computing structure, every processor can "see" all of the memory. This method is more general than typical distributed- (or fragmented-) memory computing techniques because it provides efficient access by any processor to all data without having to engage software routines on a remote processor to assist in the data transfer. More actions can be performed simultaneously, speeding execution. In addition, by letting the system conduct dynamic rescheduling—responding to run-time information—it can perform certain computations more effectively, a capability that adds to its generality. And because this arrangement is closer to the way computational scientists think about their problems, programming the system is more intuitive. Typically programmers must determine beforehand how a problem should be handled by a system, a complex and laborious task. But an HTMT system makes many of these decisions by itself, thereby helping to alleviate one of the biggest difficulties in working with large computers—programming them.

The hybrid computing system will provide greater availability to users through the use of higher-capability subcomponents, allowing it to achieve the same level of performance with fewer parts. This parts reduction increases the mean time between failures of the entire system, thus boosting operational uptime.

Another innovative aspect of the HTMT system will be its use of high-density-capacity holographic memory storage devices. This alternative to the semiconductor-based DRAM is being explored by academic and industrial research laboratories and should provide superior storage density as well as lower power consumption and costs.

Holographic storage systems use light-sensitive materials to accumulate large blocks of data. Photorefractive and spectral hole-burning techniques represent two distinct approaches. In photorefractive, storage, a plane of data modulates a laser beam (signal) that interferes with a reference beam in a small rectangular block of a storage material such as lithium niobate. The hologram results from the electro-optic effect that occurs when local electric fields are created by trapped, spatially distributed charge carriers excited by the interfering beams. Many data blocks may be stored in the same target material. They are differentiated by varying either the angle of incidence or the wavelength of the laser beam. The spectral hole-burning technique relies on a nonlinear response of a storage material to

---

Key Concepts /Hypercomputing

Contention—Time delay created when two processors try to access a shared resource simultaneously

Latency—Delay caused by the time it takes for a remote request to be serviced or for a message to travel between two processing nodes

Load Balancing—Distributing work evenly so that all processing nodes are kept occupied as the program is executed

Overhead—The time spent on noncomputational functions such as the logistic management of parallel resources and concurrent tasks

Percolation—Method of managing tasks and data movement without incurring delays caused by overhead, latency, contention or starvation

Processor-in-Memory (PIM)—Integrated circuits that contain both memory and logic of the same chip

Starvation—Wastage of computing resources caused by insufficient program parallelism or poor load balancing

Wave Division Multiplexing (WDM)—Method by which the effective bandwidth of an optical channel can be increased by using optical signals with different wavelengths

---

optical stimuli. Data are represented by changes in the photosensitive medium's absorption spectrum. Many bits can be stored at a given spatial location.

Photorefractive methods are far more advanced. But in the long-term, spectral hole-burning technology may yield significantly higher memory density. Typical holographic devices currently feature access times of several milliseconds—approximately the same as conventional secondary storage devices such as hard disks and CD-ROM drives. But advanced techniques employing tunable lasers or arrays of laser diodes each set at a slightly different angle to one another are expected to yield access times of a few tens of microseconds. Although these access times are about two orders of magnitude longer than that of DRAM, their data bandwidths are the same or greater, and the systems are about 100 times faster than conventional disk drives. Storage capacities of 10 gigabits or more in blocks as small as a few cubic centimeters are expected within the next decade.

To connect the speedy superconducting processors and high-density holographic memory systems in a network, HTMT will use high-capacity optical data pipelines. Instead of employing electrons in metal wires, HTMT will speed communications by using photons in fiber-optic cables. Wires can easily handle hundreds of megabits per second, and speeds of a few gigabits per second (gbps) can be achieved by using differential pairs of input/output pins (one goes up while the other goes down). But it could take tens of millions of wires to supply all the global communications bandwidth required of systems operating in the petaflops regime. With modulated lasers, digital light signals can transmit at up to 10 gbps per channel or more in conventional optical communications systems.

Employing multiple wavelengths (or colors) of light carrying digital information dramatically improves fiber-optic bandwidth or channel capacity. HTMT will use an advanced optical transmission system called wave division multiplex (WDM) communications. It should provide about 100 times the per-channel bandwidth of the best conventional metal-wire communications systems. WDM allows separate digital signals, each with its own dedicated light wavelength, to travel together through the same channel. The number of different wavelengths that can be simultaneously transmitted through a single channel has grown to around 100 in recent years, and in time this figure may rise further. With improved receiver, transmitter and switch technology now in development, switching rates of 50 megahertz or more will soon be possible. Still experimental devices may bring about rates on the order of one gigahertz in the future. This capacity level would be sufficient to manage the huge information flow of a petaflops-scale computing system.

These next-generation hypercomputers would offer an important tool for exploring the world's most pressing problems, including global warming, disease epidemics and cleaner energy. In 1999 the President's Information Technology Advisory Committee strongly recommended financial support for these kinds of projects. Research groups have demonstrated that HTMT technologies could be the best route to trans-petaflops performance. Proper funding is all that is needed to put these systems in place.

crochips have seamlessly in
ated into our environment. T
everywhere-in grocery check
nes, inside the workings of
meras, under the hoods of ca
ey direct the functioning of
tches, artificial hearts, an
tomotive assembly lines. Som
wly designed fighter jets ar
stable that without constant

Conclusion

M ore speed. More power. More possibilities. These are the elements that create a draw to manufacture the most powerful computers in the world.

Supercomputers of today are largely used to map and predict complex systems. Weather prediction, which to many seems to be a hit or miss proposition, is one area where the sheer processing power of supercomputers finds a natural application. Tracking and predicting the behavior of air currents over a wing surface is another.

In our drive to create ever more powerful computers, the very material that composes microprocessors is being stressed to its physical limits. The ever-shrinking silicon microchip is coming very close to the point where it will simply fail to function. What will happen when those limits are reached? Will the barriers dictated by physics stifle the growth of ever-faster computers?

Several strategies covered in this book manage to sidestep this problem and in theory could wring additional speed from existing systems, but the problem remains. Using new materials for microprocessors is one alternative solution. Dividing complex tasks into small parcels that are handled by many conventional computers is another. Yet, a fundamental shift in the nature of computers and how they operate seems to be the only solution to taking processing speed to new levels.

Today, quantum computing stands out as perhaps the most intriguing pathway to the next generation of supercomputers. What would take a present-day computer 100 steps to complete would take a quantum computer but four steps to accomplish. The jump in speed that this represents could revolutionize the nature of computers, and in turn, revolutionize how we use computers.

If the problems blocking the creation of a feasible quantum computer are overcome, it could be that in 50 years computers will look nothing like the machines we use now. In speed and power, these next-generation devices would make the most powerful computers currently on the planet appear to be as primitive as the idea, once held by many people, that the world is flat or that the earth is the center of the solar system.

Human sciences are a direct reflection of ourselves. In order for computers to evolve, the inventors of computers must evolve, too. This evolution won't appear on a physical level, but in the manner of how people think and how they perceive the world.

Index

Photo Credits

Pg. 8-9, David Fierstein. Pg. 33, Johnny Johnson. Pg. 37, IBM Corporation. Pg. 55, Joe Ziff. Pg. 85–86, Michael Goodman. Pg. 105, Dusan Petricic. Pg. 117, Samuel Velasco.